The
Reference Shelf ®

U.S. Infrastructure

Edited by

Paul McCaffrey

The Reference Shelf
Volume 83 • Number 4
H. W. Wilson
A Division of EBSCO Publishing, Inc.
Ipswich, Massachusetts
2011

The Reference Shelf

The books in this series contain reprints of articles, excerpts from books, addresses on current issues, and studies of social trends in the United States and other countries. There are six separately bound numbers in each volume, all of which are usually published in the same calendar year. Numbers one through five are each devoted to a single subject, providing background information and discussion from various points of view and concluding with a subject index and comprehensive bibliography that lists books, pamphlets, and abstracts of additional articles on the subject. The final number of each volume is a collection of recent speeches, and it contains a cumulative speaker index. Books in the series may be purchased individually or on subscription.

Library of Congress has cataloged this serial title as follows:

U.S. infrastructure / edited by Paul McCaffrey.
    p. cm. -- (The reference shelf ; 83, no. 4)
 Includes bibliographical references and index.
 ISBN 978-0-8242-1108-0 (alk. paper)
1. Infrastructure (Economics)--United States. 2. Communication and traffic--United States. 3. Transportation--United States. 4. Public utilities--United States. I. McCaffrey, Paul, 1977- II. Title: US infrastructure. III. Title: United States infrastructure.
 HC110.C3U215 2011
 363.60973--dc23
                          2011022398

Cover: The George Washington Bridge, as seen from the Washington Heights section of Manhattan, in New York City. Courtesy of Sandra Hundacker (www.hundertmorgen. net.)

Visit H. W. Wilson's Web site: www.hwwilson.com

Printed in the United States of America

Contents

Preface

At the height of the evening rush hour, at approximately 6:05 P.M. on August 1, 2007, a section of the I-35W Mississippi River Bridge in Minneapolis, Minnesota, collapsed, sending cars and trucks tumbling into the waters below. The tragedy, which investigators determined was caused by a defective gusset plate used in the span's construction, took the lives of 13 people and injured dozens more. During a 2005 inspection, for reasons unrelated to the faulty gusset plate, the bridge had been deemed "structurally deficient." Nevertheless, it was not scheduled to be replaced until 2020.

Just two weeks earlier, on July 18, an underground steam pipe burst outside of New York City's Grand Central Station in Midtown Manhattan. A towering geyser of boiling mud and vapor shot into the air, raining chunks of debris and scalding scores of frightened pedestrians as they scrambled for safety. Nearly 50 bystanders were hurt during the incident and one suffered a fatal heart attack. First installed in 1924, the pipe had been declared safe by inspectors just a few hours before it exploded.

Hurricane Katrina made landfall off the coast of Louisiana on August 29, 2005. At first it looked as though the city of New Orleans had avoided catastrophe. Rather than hitting the densely populated and low-lying Big Easy head on, Katrina delivered a glancing blow—albeit one of hurricane force. Relief was fleeting. The powerful storm surge churned up by Katrina breached a number of the city's levees and water came gushing through the gaps. A few days later, most of New Orleans was submerged. Designed by the U.S. Army Corps of Engineers to save the city from just such a fate, the New Orleans levee system had failed in spectacular fashion and helped initiate one of the greatest disasters in American history.

In the face of such mishaps, many have begun to wonder whether the nation's infrastructure—specifically its transit, energy, and communication networks, water management resources, and other facilities critical to its functioning—is in crisis. These concerns are both immediate and long-term. On one hand, as recent disasters demonstrate, crumbling roads and bridges and unmaintained or deficient dams, pipes, and levees pose an ongoing threat to human life. But even if all our deteriorating infrastructure were repaired overnight, there would still be cause for concern.

As the world grows ever more globalized and interconnected, economic competition—for technology, markets, and diminishing natural resources—is expected to increase. Further complicating this picture is the threat of global climate change—a phenomenon that many believe will radically shape the contours of the future in ways both anticipated and unforeseen. How we adapt our infrastructure to meet these twin challenges will go a long way in determining how we emerge from them. It stands to reason that the nation with the best infrastructure, that develops the most up-to-date transit grid, fosters the best power and communications networks, and manages its energy consumption and natural resources most efficiently, will have a distinct advantage in the years ahead.

Still, in the United States there is vast disagreement as to what effective infrastructure looks like. For example, consider our transportation grid. Whether going to school, work, or the store, most Americans first climb into some sort of automobile. Yet anyone who has sat for hours in gridlocked traffic, emptied their wallet to fill a gas tank, or suffered from breathing smoggy air knows that the system is far from perfect. Indeed, each year there are over 30,000 fatal automobile accidents. Despite its drawbacks, the car is a long way from falling out of favor. It has come to symbolize American freedom and mobility, allowing people to live where they want and come and go on their own timetable. As a consequence, we are far from developing a viable alternative.

Another infrastructure impediment in the United States is deciding just who should build it. A suspicion of government is ingrained in the American psyche, so many are reluctant to let public agencies take the lead. Of course, others are similarly wary of leaving such essential services in the hands of private, for-profit companies.

Divided into five chapters, this volume illuminates some of the major issues confronting U.S. transport, water, and power and communications facilities and how they might be best addressed. Selections in the first chapter, "Breached Levees, Fallen Bridges: Is American Infrastructure in Crisis?" offer a broad overview of the subject, noting the specific areas of concern, the dangers they pose, and their potential solutions. The succeeding chapters divide American infrastructure into its component parts, taking an in-depth look at each. In the second section, "Highways, Byways, and Railways: Transportation Infrastructure," articles consider the state of the American transit system. Selections in the subsequent chapter, "The Train Debate: Can Rail Revolutionize American Transport?" continue the exploration of U.S. transit, examining the potential for rail, both high-speed and otherwise, to transform American mobility. Long a dream of environmentalists, expanded rail transport must overcome a number of obstacles if it is to achieve the sort of prominence its proponents envision. In "The Grid: Power and Communications Networks," entries discuss energy and communications infrastructure. In the final section, "From Taps to Toilets: Waterworks," articles consider the state of American water facilities.

In conclusion, I would like to thank the many authors and publishers who allowed us to reprint their work in these pages. I'd also like to thank the many friends and colleagues here at the H.W. Wilson Company who helped in putting this volume together, especially Joseph Miller, Ken Partridge, Rich Stein, and Carolyn Ellis.

Paul McCaffrey
August 2011

1

Breached Levees, Fallen Bridges:
Is American Infrastructure in Crisis?

A destroyed tow truck sits in a hole in the ground at the site of an underground steam pipe explosion on July 18, 2007, in New York City. The explosion tore a crater in Lexington Avenue near Grand Central Terminal, sending residents running for cover amid a towering geyser of steam and leaving asbestos in the dust that settled, but city officials said tests indicated the air was safe of the carcinogen.

Cars rest on the collapsed portion of I-35W Mississippi River Bridge, after the collapse on August 1, 2007. This was featured as one of the 12 most powerful photos of 2007 on ABC News on-line.

Editor's Introduction

In January 2009, the American Society of Civil Engineers (ASCE) issued its Report Card for America's Infrastructure. In compiling the grades, ASCE analyzed the conditions of a range of the nation's critical facilities, from roads and bridges to dams and levees. The results were deeply troubling. No aspect of U.S. infrastructure warranted a grade higher than a C+, while such vital components as drinking water, wastewater treatment and disposal, levees, roads, and inland waterways each earned a D-. In all, the combined grade point average (GPA) for all American infrastructure was a disappointing D. According to ASCE calculations, an investment of $2.2 trillion dollars over the next five years was required to bring these facilities up to speed.

While President Barack Obama's national stimulus package, the American Recovery and Reinvestment Act of 2009, allotted considerable money towards infrastructure improvements, it did not approach the $2.2 trillion figure called for by ASCE, and given record budget deficits and a difficult economic picture, the necessary funding will not materialize anytime soon. As a consequence, ASCE's goals are likely to remain unmet for the foreseeable future.

Of course, not everyone agrees with ASCE's grim assessment, and while few would claim that our infrastructure is in peak condition, many believe the dire warnings about an infrastructure crisis are overblown. The articles in this chapter attempt to ascertain just what sort of condition our critical facilities are in, noting which sectors are especially in need of revitalization and how such ends might be achieved.

For Burt Solomon, the author of the first selection, "The Real Infrastructure Crisis," the problem with our public facilities is not their physical condition, but rather an overall lack of investment. Our infrastructure functions adequately for the most part, he observes, but we are not building for the future. Since the 1980s, he contends, the American taxpayer has been reluctant to pay for improvements while infrastructure spending itself has taken on increasingly political overtones. We need to choose "between a world-class infrastructure and muddling through," Solomon says, adding that the decision will be made in what he calls "the political marketplace." "If Americans get disgusted enough, they'll do what it takes," he concludes. "Otherwise, they won't."

In the congressional elections of 2010, the Republican Party won control of the U.S. House of Representatives and critical governorships across the country. In "Reverse Gears: A New Reality for Public Works?" John McCarron discusses how these outcomes are likely to affect infrastructure funding and various public works projects. A foreshadowing of this change of gears occurred shortly before the elections, in the state of New Jersey, when Republican governor Chris Christie pulled the plug on the planned construction of a new Hudson River tunnel connecting New Jersey to New York City. Since the election, McCarron points out, Republicans governors in Ohio and Wisconsin refused hundreds of millions of dollars in federal funding for passenger rail construction. Whether wrongheaded or not, McCarron notes, these decisions represent a sea change in infrastructure planning. As Steve Elkins, a city councilman from Bloomington, Minnesota, observes, "The building binge is over. Now our mantra has got to be 'fix it first.'"

In the subsequent entry, "Look Out Below! America's Infrastructure Is Crumbling," Eric Kelderman reports on some alarming statistics: more than a quarter of U.S. bridges are obsolete or require major repairs; one in three major thoroughfares are in poor condition; failing sewer systems annually spill around 1.26 trillion gallons of raw sewage into the environment; while an increasing number of dams are in danger of collapsing. Such figures lead Donald F. Kettl of the University of Pennsylvania to observe, "Much of America is held together by Scotch tape, bailing wire and prayers."

In "American Collapse," Sarah Goldhagen attributes what she perceives as our increasingly substandard infrastructure to two principle causes. First, our settlement and transport patterns have evolved from the "city-suburb-exurb-farmland" model to one dominated by metropolitan regions. The earlier setup could be adequately managed by the federal-state-local government structures that were in place. Metropolitan regions are not so easily organized, however, as they often stretch across several states and numerous distinct municipalities, making coordination and long-term planning difficult. The second reason, Goldhagen observes, is that the federal government has increasingly delegated infrastructure responsibility to the states, which often lack the funds and the will to invest in the ambitious projects needed to build 21st century facilities.

Jonathan Masters considers what the next steps in infrastructure development ought to be by speaking with four experts: Robert Puentes, a Senior Fellow at the Brookings Institution; Stephen Goldsmith, New York City Deputy Mayor for Operations; Richard Little, Director of the Keston Institute for Public Finance and Infrastructure Policy; and Felix G. Rohatyn, a former investment banker and U.S. ambassador to France.

Erik Sofge, in "Rebuilding America Special Report: How to Fix American Infrastructure," discusses an investigation into the state of the nation's infrastructure by *Popular Mechanics*. While noting the many shortcomings illuminated by the inquiry, Sofge also highlights some innovative solutions that might be put to use to address them.

In the final piece, "A New Bank to Save Our Infrastructure," Rohatyn and Everett Ehrlich discuss the proposals of the Commission on Public Infrastructure at the Center for Strategic and International Studies (CSIS). The main initiative put forward by the Commission is one Rohatyn discusses earlier in this chapter: the establishment of a National Infrastructure Bank, which would act as "a private investment bank . . . that evaluates project proposals and assembles a portfolio of investments to pay for them."

The Real Infrastructure Crisis[*]

By Burt Solomon
National Journal, July 5, 2008

It's a frighteningly familiar catastrophe to imagine. An earthquake in Northern California ruptures 30 levees along the converging Sacramento and San Joaquin rivers, and 300 billion gallons of saltwater rush inland from San Francisco Bay, flooding 16 islands and ruining the supply of fresh water across two-thirds of the nation's most populous state. Or picture this: In southern Kentucky, the 55-year-old Wolf Creek Dam (where water has seeped through the foundations for years) gives way. The breach lets loose the largest man-made reservoir east of the Mississippi River, flooding the communities along the Cumberland River and shorting out the electric guitars in Nashville.

These were the top two horror stories—"5 Disasters Coming Soon If We Don't Rebuild U.S. Infrastructure"—that *Popular Mechanics* conjured up for its readers last fall, after the collapse of a bridge in Minnesota killed 13 innocents on their way home from work. The stunning sight of an interstate highway plunging into the Mississippi River, just two weeks after a steam pipe exploded beneath Lexington Avenue in Midtown Manhattan—and less than two years after Hurricane Katrina brought New Orleans to its knees—dramatically brought the nation's fallible infrastructure to the public's attention. So, too, did the overwhelmed levees along the Midwestern rivers during the recent rains. And so did the garden-variety failures, such as the water main break on June 16 in Montgomery County, Md., bordering Washington that forced some of the capital's bigwigs to boil water before brushing their teeth.

In the mammoth but aging networks of roads, bridges, railroads, air traffic, sewers, pipelines, supplies of fresh water, and electricity grids that helped turn the United States into the world's economic superpower, other dangers lurk. All over the country, clean-water and wastewater facilities are wearing out. The combined sewers that 40 million people in 772 cities use could disgorge their raw contents into waterways when the next storm passes through. Every summer brings the possibility of blackouts.

Traffic gridlock has become a fact of life, jamming the highways and airways and creating bottlenecks of goods through the ports, especially around Los Angeles and New York City. The American Society of Civil Engineers has classified 3,500 of the nation's 79,000 dams as unsafe; in a 2005 report card on the nation's infrastructure, the society assigned grades that ranged from C+ (for the proper disposal of solid waste) down to D (for the supply of drinking water and the treatment of wastewater).

Talk of the "crisis" in the nation's physical infrastructure has leapt beyond think-tank forums and earnest editorials. It has quickened legislators' interest, generated heartfelt lobbying on Capitol Hill—expected to climax next year when Congress must reauthorize the pork-laden highway program—and nosed its way into the presidential campaign.

Experts, however, consider "crisis" an overblown description of the perils that America's infrastructure poses. Federal investigators have tentatively concluded that the ill-fated Interstate 35W in Minnesota collapsed not because it was structurally deficient—although it was—but because of a design defect: The gusset plates connecting the steel beams were half as thick as they should have been. Nationwide, bridges are in better structural condition than they were 20 years ago, and the most critical of the nation's 4 million miles of roadways are in pretty good shape. In the transportation system, "the physical condition has not noticeably deteriorated . . . in the past two decades," said Katherine A. Siggerud, the managing director of physical infrastructure issues at the determinedly nonpolitical Government Accountability Office. "The condition of the most-traveled roads and bridges in the United States, the interstates and the national highways, [has] improved in quality."

The more serious problem is the lack of roads and the traffic congestion that this shortage creates, especially around major cities. In the nation's airways, too, congestion has become chronic, especially at airports in the Northeast. But Gerald Dillingham, the GAO's director of civil aviation issues, doesn't see a crisis in the near or midterm, and he is hopeful that better technology and new ways of structuring the airways can stave off disaster for at least the next 15 years. The Transportation Department has calculated the overall economic cost of congestion at $200 billion a year, surely a drag on the nation's commerce, not to mention a vexation to anyone stuck in traffic. Still, in a $14 trillion economy, that amounts to 1.4 percent—a pittance.

Fixing the nation's infrastructure is "a matter of fine-tuning the economic production system," said Kenneth A. Small, an economist who specializes in transportation at the University of California (Irvine), "not a matter of moral outrage." Rudolph G. Penner, a senior fellow at the Urban Institute, said, "I'd call it a problem, not a crisis." Even the lobbyists who urge more spending on the nation's infrastructure acknowledge that the assertions of impending doom are an exaggeration. Janet F. Kavinoky, the director of transportation infrastructure at the U.S. Chamber of

Commerce, is the executive director of Americans for Transportation Mobility, an alliance of construction companies and labor unions. "If you don't say it's a crisis," she explained, "nobody shows up at your press conference."

Nor is the country ignoring the issue. The nation's spending on infrastructure continues to rise; New Orleans is rebuilding the levees that Katrina breached. "The things that need to get done are getting done, by and large," said Timothy P. Lynch, the American Trucking Associations' senior vice president for federal relations and strategic planning.

This isn't to say, of course, that all is hunky-dory. The future of U.S. infrastructure could be grim indeed if too little is done. At the core, it's a question of cost. Bridges and roads are expensive—to build or to fix—and so are mass transit, airport runways, and almost everything else. The civil engineers issued a widely invoked price tag of $1.6 trillion over five years to do what needs to be done, but even champions of a strong infrastructure find such a number inflated—"a compilation of a wish list," the ATA's Lynch said.

Moreover, investment bankers say that plenty of capital is available for work that is critical to the nation's well-being. What may be missing, however, is the political will to spend this capital. Increasingly, legislators and local governments are trying to arrange infrastructure financing in ways that conceal the true costs from taxpayers, who are reluctant to foot the bill, and that may transfer the financial burdens to future generations. If the measure of a society's responsibility is its willingness to invest for the long run, then the crisis in infrastructure is this: Do Americans possess the national will to pay for what their children and their children's children are going to need?

ANCIENT ROME MEETS REAGAN

Only occasionally has a civilization made its infrastructure an emblem of its ambition or greatness. Consider, notably, the marvels of ancient Rome—its roads, its aqueducts, its public baths and lavatories, its Colosseum and other sites of public entertainment. Conceived as a military necessity to assure the movement of troops through a far-flung empire, Rome's extravagant and enduring infrastructure took on other functions, too. As a public benefaction, it gave the state a way to justify its own existence, according to Garrett G. Fagan, a historian at Pennsylvania State University, and the many amenities that wealthy families financed served as "a kind of social compact between the upper classes and the poorer classes." The boldness and breadth of Roman infrastructure, Fagan said, "go a long way to explain why the empire lasted so long."

The United States has often shown a similar ambition. In 1808, after Thomas Jefferson's Louisiana Purchase added a vast wilderness that stretched as far as the future Montana, Treasury Secretary Albert Gallatin proposed a national transportation network of roads, rivers, and ports.

In the following decades, Henry Clay of Kentucky lent his legislative weight in the House, and then in the Senate, to the "internal improvements" of canals and railroads. Abraham Lincoln, even as he struggled to win the Civil War, pursued plans for a transcontinental railroad. Theodore Roosevelt, so fond of proclaiming the needs of "future generations," convened a conference of governors that resulted in water projects that irrigated the West and generated electricity cheaply; his list of ventures-still-undone gave TR's fifth cousin, Franklin D. Roosevelt, a starting point when he tried to spend the nation out of the Great Depression. Then, in the postwar boom of the 1950s, President Eisenhower pressed for a system of interstate highways that knitted the nation together and bolstered its economy. As late as the 1970s, after the Cuyahoga River in Cleveland caught fire in 1969, the federal government invested tens of billions of dollars in sewer systems and wastewater treatment plants.

Taxpayers' generosity toward the nation's infrastructure, however, took a dive during the 1980s. President Reagan's aversion to using taxes for domestic spending, exacerbated by Wall Street's obsession with quarterly earnings, encouraged a shortsightedness in assessing the public good. According to Sherle R. Schwenninger, the director of the New America Foundation's economic growth program, the money that government at all levels has devoted to infrastructure, as a proportion of the nation's total economic output, slipped from 3 percent during the 1950s and 1960s to only 2 percent in recent years.

"We've just not reinvested," former Council of Economic Advisers Chairman Martin N. Baily complained at a Brookings Institution forum last fall, "because nobody wanted to raise the taxes to do that." Even in Katrina-devastated Louisiana, when the Army Corps of Engineers announced in 2006 that its estimate for fixing the levees had ballooned from $3.5 billion to $9.5 billion, the state's politicians and editorial writers wailed.

NOT TO WORRY

It wouldn't take many years, or so it is said, before the weeds poked up through a neglected interstate highway. Not to worry. Even as the nation's enthusiasm for long-term investments has flagged, the total amount of money spent on its infrastructure has continued to grow. As the federal share has shrunk (from 32 percent in 1982 to less than 24 percent in 2004, according to the Congressional Budget Office), state and local governments have picked up the slack. Counting all levels of government, public entities spent $312 billion on the nation's transportation and water infrastructure in 2004, three times as much—after taking inflation into account—as in 1956, when Eisenhower's heyday began.

Has the U.S. underfunded its infrastructure, on which its economy rests? "Compared to what we really need, I think so," said Penner, a former CBO director, "but relatively slightly."

Consider, for example, the state of the nation's bridges. Last summer's tragedy in Minnesota cast a spotlight on the Federal Highway Administration's alarming conclusion that, as of last December, 12 percent of the nation's bridges were structurally deficient. But less attention was paid to the fact that this proportion had shrunk from 13 percent in 2004 and nearly 19 percent in 1994. Nor was it widely noticed that the label of "structurally deficient" covered a range of poor conditions, from serious to far less so. Fewer than a tenth of the tens of thousands of bridges deemed deficient are anywhere close to falling down. (A Federal Highway Administration spokeswomen said the agency does not have summary information about the location and size of the worst bridges.)

The surge of bridge inspections that followed the disaster in Minnesota turned up a second bridge with bowed gusset plates across the Mississippi in Minneapolis-St. Paul—it was immediately closed and slated for repairs—and another one in Duluth. The Minnesota Legislature found numerous shortcomings in the state inspectors' work on the I35 bridge that had been tagged as structurally deficient for some cracking and fatigue. According to the National Transportation Safety Board's investigators, however, the inspectors were not the problem. Indeed, the investigators cited the effort to repair the bridge, which entailed piling construction supplies and equipment on its overburdened deck, and the thin gusset plates as the likely leading causes of the I35 collapse. The more that they have learned about the disaster, the less it has served as a morality tale.

As for a fear of falling bridges, "I don't really think we're in a crisis," said economist Small. He also mentioned the "pretty strong" system of bridge inspections and placed the 13 deaths in Minnesota into the context of all U.S. traffic fatalities, which average 120 a day. "If you plot the statistics," he noted, "you might not notice the bump."

On the roads, too, drunk drivers or malfunctioning vehicles cause many more deaths than potholes or crumbling concrete. The roads are OK, but there aren't enough of them to hold the traffic, and building more will only increase demand. The gridlock is worst of all around Los Angeles, the San Francisco Bay area, Chicago, New York City, Atlanta, and Washington, but it has also spread into unlikelier venues. A third lane is being built along certain truck-clogged stretches of Interstate 80 in Iowa and Nebraska. The GAO's Siggerud pointed to "bottlenecks in every mode of transportation," which stand to get worse. The Federal Aviation Administration has predicted that air traffic may triple during the next two decades, and the American Road & Transportation Builders Association has forecast that the volume of cargo on U.S. roads will double. In Los Angeles, the freight volume is expected to triple as the population grows by 60 percent, producing strains that the U.S. chamber's Kavinoky warned "will paralyze the city."

Ian Grossman, the FHWA's associate administrator for public affairs, lamented the Little League games unattended and the volunteerism in decline because of congestion. "It shouldn't be a fact of life," he said.

The economic impact of the bottlenecks has been "woefully understudied," according to Robert Puentes, an expert in infrastructure at the Brookings Institution, who regards transportation policy as "a fact-free zone." But Clifford Winston, an economist at Brookings, has tried. His calculation of the annual economic cost of congestion is just a third of DOT's—$15 billion in air traffic and nearly $50 billion on the roads, counting the shipping delays, the higher inventories required, the wasted fuel, the value of gridlocked motorists' time, and other not-quite-tangible factors. The impediments are numerous, Winston said, but "none of them are big. That's why they persist."

The problem of congestion is, to a degree, self-limiting. It could injure the economy of a gridlocked metropolis, but by no more than 5 to 10 percent, according to Small, by driving business to the suburbs, exurbs, and smaller cities that stand to benefit from the big cities' pain.

Nor has congestion in the air been neglected. The air traffic system, in which 25 percent of last year's flights arrived late, has added runways in recent years in Atlanta, Boston, Cincinnati, Minneapolis, and St. Louis; starting this November it will add another runway at Chicago's O'Hare. The $13 billion that the FAA spends annually on infrastructure development for civil aviation falls a mere $1 billion short—pocket change, really—of what GAO analyst Dillingham believes it should spend. The next generation of air traffic control, based on a global positioning system instead of on radar, has been delayed—not because of the immense cost or the technology, Dillingham said, but because of the difficulty of integrating it into the existing system.

Scarier, perhaps, for the nation's economic future is the possibility that congestion or other strains on an elderly infrastructure will damage America's already shaky competitive position in global markets. The American business executives who leave South Korea's luxurious Incheon International Airport or Shanghai's modern, half-empty airport to arrive at New York's seedy JFK are bound to feel repulsed. Today, that is nothing more than inconvenience, but eventually, economists say, it could count.

"In a globalized economy," the New America Foundation's Schwenninger said, "there are only a few ways you can compete." Asian countries can claim lower wage rates and taxes, and Europe boasts governmental subsidies and an educated workforce. This leaves infrastructure, Schwenninger ventured, as American businesses' best hope for a competitive edge—more so than 20 to 30 years ago, and more important than education. Silicon Valley, he reported, has lost some of its silicon-wafer manufacturing to Texas and countries overseas because producers fear brownouts in California.

Yet the threat to U.S. competitiveness shouldn't be exaggerated, for other countries face similar problems with congestion. Gaining permission to build a new road or runway is even harder in cramped, environmentally conscious Europe. China and India are spending 9 percent and 5 percent, respectively, of their gross domestic product on infrastructure. The U.S., however, has an overwhelming advantage: Its elaborate infrastructure—4 million miles of roads, 600,000 bridges, 26,000 miles of commercially navigable waterways, 11,000 miles of transit lines,

500 train stations, 300 ports, 19,000 airports, 55,000 community drinking water systems, and 30,000 wastewater plants—is already built.

DUCKING THE COSTS

Still, on matters of infrastructure, the United States is losing ground. "It would be an overstatement to say our system is in crisis," Brookings's Winston said. "At the same time, the annual costs of the inefficiencies [because of congestion] are large, growing, and unlikely to be addressed by the public sector."

No longer is American infrastructure on the cutting edge. "I think we are falling behind the rest of the world," Rep. Earl Blumenauer, an earnest veteran Democrat from Portland, Ore., said in an interview. He is pushing legislation to create a blue-ribbon commission that would frame a coherent national vision for dealing with the country's disparately owned and operated infrastructure, variously the responsibility of federal, state, or local governments or—for a majority of dams and many recent water systems—private owners. Besides the existing bottlenecks in the movement of goods, Blumenauer foresees "real problems with the backlog of projects"—for sewers, roads, water, bridges, etc.—within five to 10 years. And deferring maintenance, he noted, increases the costs, which is one reason he thinks that the astronomical price tags "tend to be understated, not overstated."

The GAO, among others, is more skeptical, not only of the civil engineers' $1.6 trillion, $300-billion-plus-a-year cost projection but also of a congressionally created panel's recommendations. The National Surface Transportation Policy and Revenue Study Commission announced in January that the nation must spend $225 billion annually—$140 billion more than at present—on its roads, waterways, and railroads. "Most of the needs assessments," the Urban Institute's Penner explained, "are very much influenced by special interests," using unrealistic assumptions and self-serving estimates.

How much the nation must spend, however, is certain to rise. For fresh water and wastewater alone, by the GAO's calculations, the infrastructure costs over the next 20 years will range between $400 billion and nearly $1.2 trillion to correct past underinvestment. The existing facilities, if not repaired or replaced, would probably take 10 to 20 years to deteriorate, an offical said, not two or three.

Given the presumed reluctance of American taxpayers to pay up front, such projections have quickened the search for politically palatable alternatives to financing infrastructure projects—artful ways of ducking the costs. Hence the rising popularity of public-private partnerships, "or as we called them, business deals," Everett M. Ehrlich, an expert on infrastructure financing, told the House Transportation and Infrastructure Committee in June. On May 19, Pennsylvania Gov. Ed Rendell had announced the winning $12.8 billion bid (submitted by a Spanish toll-road company and a division of Citigroup) for a 75-year lease of the Pennsylvania Turnpike. The idea wasn't original. The city of Chicago signed a $1.8 billion lease for the Chicago Skyway in 2005 and has received a half-dozen bids for privatizing Midway Airport. The Indiana Toll Road was leased in 2006 for $3.8 billion. A private company built and runs the Dulles Toll Road in Northern Virginia, and the Texas Legislature has imposed a two-year moratorium on a planned network of private toll roads out of concern that the deals were too lucrative for the operators.

A private operator, the thinking goes, can raise tolls with an abandon that would give politicians the willies, and investment banks are salivating at the prospect of jumping in. But the criticism has mounted. "Deferred maintenance will become a big part of creating profits for shareholders," Allen Zimmerman, a resident of South Whitehall Township, Pa., warned about leasing the turnpike, in a letter to The Morning Call of Allentown, Pa. Economists worry that a private operator might milk the drivers along the popular routes while ignoring the boondocks.

By the GAO's lights, the value of any given deal depends on the particulars, such as the quality of the management, the assurances of proper maintenance, and the uses to which a state will put the newfound revenues. Indiana is spending its bump in revenue on a 10-year transportation plan; Chicago, on the other hand, has pointedly refrained from any allocation. Pennsylvania officials have vowed to spend their windfall on transportation but have been "evasive," the ATA's Lynch said, about specifics.

At least so far, the greatest hindrance to an influx of private capital for the nation's infrastructure, according to Penner of the Urban Institute, is the paucity of investment opportunities. He also lacks faith in the other ideas being pitched on Capitol Hill that seek to lure capital while dodging the costs—notably, proposals to establish an infrastructure "bank" to leverage private investments and to institute a separate capital budget for the federal government. Nor does the direct approach—the possibility of federal appropriations—give him reason to hope. He fears that the entitlement programs (Social Security, Medicare, and Medicaid) will squeeze the budget, shrinking the discretionary spending on infrastructure projects.

Where, then, will the money come from? At Brookings, infrastructure expert Puentes thinks that relatively small, targeted investments can relieve the worst bottlenecks—those of national importance, such as the congestion at the port of Long Beach, Calif. In any event, simply relying on the construction of new highways and airport terminals won't suffice, in Small's view: "It's just too expensive."

Many economists favor another solution—congestion pricing. London, Stockholm, and Singapore now charge vehicles that drive into the central cities at busy times of the day. Michael Bloomberg, New York City's business-man-turned-mayor, pursued the idea until the state legislature shot it down. Pure congestion pricing, a high-tech means of raising or lowering the toll depending on the traffic, is being tested on a highway north of San Diego, where the price of driving changes every few minutes. Such pricing would be one way for Americans to pay their way.

THE POLITICAL MARKETPLACE

On the night that Sen. Barack Obama of Illinois claimed the Democratic nomination for the presidency, he spoke to the nation about, among a litany of intentions, "investing in our crumbling infrastructure." Of course, he happened to be in Minnesota, less than 10 miles from where the I35 bridge had collapsed. But then he spoke of the problem again two days later while campaigning in Virginia and, later, at a roundtable with 16 Democratic governors. In trying to bolster his appeal to working-class voters in Flint, Mich., on June 16, Obama promised to use the money he would save from ending the war in Iraq on a National Infrastructure Reinvestment Bank that would spend $60 billion over 10 years. Stressing the issue helps Obama look sober and serious about the nation's long-term needs, which is useful for a candidate who is criticized for being inexperienced.

Sen. Hillary Rodham Clinton of New York, whom Obama bested for the nomination, demonstrated the versatility of infrastructure as a political issue. A week after the bridge fell in Minnesota, she delivered a speech in New Hampshire on infrastructure as "a silent crisis." She showed a thorough understanding of the issue ("Today nearly half the locks on our waterways are obsolete.") and offered a detailed plan of attack, including a $10 billion emergency repair fund, $1.5 billion for public transit, $1 billion for intercity passenger railways, and sundry other millions for additional projects. Nine months later, however, facing political death as Indiana Democrats readied to vote, she climbed onto the back of a pickup truck and appealed to voters beleaguered by the soaring price of fuel. Her idea? Suspend the federal gasoline tax, which pays for the upkeep on the nation's pivotal highways. Economists gagged at the thought, but Indiana Democrats rewarded her with a narrow victory.

The presumed Republican nominee, Sen. John McCain of Arizona, who agreed with Clinton on the gasoline tax, has also used infrastructure as a political football. It's a word he reveres. In campaign speeches, he has applied "infrastructure" to public health, alternative fuels, "the infrastructure of civil society," and "the Republican infrastructure." But in the conventional sense, he has linked it to one of his trademark issues. "The problem with roads and infrastructure and bridges and tunnels in America can be laid right at the doorstep of Congress," he said in May, four months after federal investigators blamed the Minnesota bridge collapse on a design flaw, "because the pork-barrel, earmark spending, such as the 'Bridge to Nowhere' in Alaska, has diverted people's hard-earned tax dollars that they pay at the gas pump." This charge drew a public rebuke from Tim Pawlenty, Minnesota's Republican governor, who is a national cochairman of McCain's campaign and is often mentioned as a possible running mate. "I don't know what he's basing that on," Pawlenty said, "other than the general premise that projects got misprioritized throughout time."

One legislator's pork, of course, is another's infrastructure. Such criticism of "pork," as a result, has not dampened Congress's enthusiasm for spending money on highways and such. The 2005 highway legislation (known, improbably, as SAFETEA-LU) authorized $286 billion over six years, $32 billion more than the Bush administration wanted. But this amount was miserly compared with the House-approved $380 billion. Members of Congress earmarked just one-tenth of the money for particular projects back home, and not all of those were considered boondoggles. An earmark, for instance, funded the newly built Woodrow Wilson Bridge along the Capital Beltway between Virginia and Maryland.

The lobbyists for the labor unions and the contractors that stand to benefit from road construction are already gearing up for next year's effort to reauthorize the highway bill. The pot will surely grow bigger—reportedly to $500 billion over six years—especially if a Democratic president works with a Democratic Congress. Spending on infrastructure has recently been touted by Rockefeller Foundation President Judith Rodin, among others, as a Keynesian response to an impending recession. And even if earmark-happy highway bills inevitably waste money, they may be worthy of praise for paying up front for whichever roads and bridges—to nowhere or to somewhere—the democratic system has deemed worthy. "In the end, there's no substitute for making systematic investment," Rep. Blumenauer said.

As a political issue, infrastructure is the kind that democracies have a hard time with—a chronic, usually invisible problem that only occasionally becomes acute. For better or worse, however, politics has become inseparable from the battles over infrastructure, sometimes to the point of amusement. When members of the House Transportation and Infrastructure Committee discussed the fateful gusset plates in Minnesota, the Republicans stressed the arbitrary nature of such a failure, which money would never have averted, while the Democrats kept mentioning the bridge's wear and tear, for which more money would have mattered. Partisan positions on gusset plates—who knew?

Still, the politics of infrastructure are far from straightforward. Earmarks and pork find enthusiasts and critics within both political parties. Congestion pricing has produced odd bedfellows. Both Bush administration conservatives and environmental activists approve of such a market mechanism that would save fuel and improve economic efficiency, while some Democrats worry about the effect of "Lexus lanes" on the poor.

The true political divide may lie between Americans who'll be willing and able to pay up front for the nation's needs—whether through taxes or tolls—and those who would rather skimp or burden their children. This sort of decision, between a world-class infrastructure and muddling through, will be made in the political marketplace. If Americans get disgusted enough, they'll do what it takes. Otherwise, they won't.

Reverse Gears: A New Reality for Public Works?*

What the November Elections Might Mean for Infrastructure Spending

By John McCarron
Planning, February 2011

Just when it looked like America had made up its mind about smarter and time-lier investments in public infrastructure, a not-so-funny thing happened on our way to a promised land of rust-free bridges and perfect, pothole-free streets.

Last November's midterm elections and the Republican capture of the U.S. House of Representatives—not to mention several key governorships and state-houses—has triggered a dramatic rethinking of the way the nation pays for public works, from sexy intercity bullet trains to prosaic local sewers.

To be sure, some kind of reckoning was probably overdue, election or no elec-tion. Infrastructure funding just isn't there like it used to be. It's one thing for spending advocates like the American Society of Civil Engineers to issue a grade of "D" and warn that government needs to invest $2.2 trillion in the nation's infrastructure over the next five years. It's another to reconcile that advice with recommendations delivered by President Obama's bipartisan commission on fiscal reform. The National Commission on Fiscal Responsibility and Reform reported in December that Washington needs to slash spending by $4 trillion by decade's end if it is to have any chance of getting deficits back in line with gross domestic product without reneging on entitlements like Social Security and Medicare.

Yet that GDP won't grow properly, according to the infrastructure advocates, if we keep starving the skeleton and arterial system on which our economy moves. Think traffic jams, freight rail bottlenecks, and silted-in intermodal harbors. The ASCE places our annual cost of congestion at more than $85 billion. And that doesn't include lives that will be lost as our bridges continue to collapse (think Minneapolis), our dams and levees breach (New Orleans), and our planes collide. The latter is a potential consequence of an air traffic control system that has yet to install a guidance system the likes of which can be had at any new car dealership.

This debate may sound familiar, but last November's election changed its dynamic, forcing infra-advocates to go on the defensive rather than push for nine-figure spending increases. Some of Washington's new legislators are especially down on earmarks—the designation of specific projects in specific locales that is the political grease of appropriation bills. Congress isn't likely to officially forgo them entirely, but the sentiment behind it cannot be ignored.

GOODBYE, JIM OBERSTAR

The signal event for planners in the big turnaround, however, was the electoral defeat of Rep. James Oberstar (D-Minn.), the powerful chair of the House Transportation and Infrastructure Committee. Representing the iron and grain funnel that is Duluth, Oberstar "got" the linkage between transportation and economic growth. The 18-term congressman also was credited with opening the appropriation process to nontraditional modes such as bikeways, light rail, bus rapid transit, and even incentives for coordinated regional planning.

Yet during his final term Oberstar lost his battle to craft a six-year, $500 billion transportation reauthorization bill, a failure that forced Congress to pass a series of temporary extensions to the expired SAFETEA-LU spending plan just to keep federal matching dollars flowing. The main hangup: how and how much to increase the federal gasoline tax.

The gas tax, the primary source of federal transportation funding, was last raised to 18.4 cents a gallon in 1993. Its yield has been declining in real dollars ever since, what with creeping inflation and improving fuel efficiency. Suffice it to say that, with a White House preoccupied by health care and a Congress anxious about unemployment and deficit spending, reauthorization with a needed gas tax hike simply didn't happen before last November's election.

Incoming committee chairman Rep. John Mica (R-Fla.) early on declared that a gas tax increase was "off the table"—a sentiment shared, at least for now, by President Barack Obama out of concern for its impact on the frail economic recovery. But Mica has talked about passing a multiyear transportation reauthorization bill "early this year," hinting that additional road and transit funds could be had by tapping "billions in special accounts." He specifically mentioned $6 billion sitting in a harbor maintenance trust.

NOT SO SHOVEL-READY

All the shifting about in Washington has played hob, of course, with infrastructure planning and execution at the state and local levels. Federal matching funds typically underwrite at least half—often 80 percent—of major public works undertakings. So while states and municipalities have their wish lists (aka transportation improvement plans, or TIPs) they are hesitant to order necessary-but-pricey engineering and environmental studies without first having their federal partner in place.

This helps to explain the lack of "shovel-ready" projects eligible for the $27 billion made available to states for highway upgrades early in the Obama administration as part of its $787 billion American Recovery and Reinvestment Act. An orgy of 14,000 road widenings, repavings, and resignalizations was to create or save 150,000 jobs—and perhaps it still will—but local prep work has taken longer than expected and by the end of 2010 only two-thirds of that target was reached, according to Federal Highway Administration data.

TOP 10 STATES' PER CAPITA EARMARKS	
State	**Per capita**
Hawaii	$318
North Dakota	$234
West Virginia	$174
Vermont	$161
Mississippi	$142
Alaska	$140
Montana	$125
South Dakota	$112
Nevada	$79
Rhode Island	$79

Source: Taxpayers for Common Sense FY 2010

Further, a growing political movement questions the fundamental wisdom of relying on public works as an economic booster shot. This attitude is new. The works-jobs linkage has been an article of faith in American governance since the New Deal. Every schoolchild is taught how President Franklin Roosevelt put millions of idle Americans to work during the Great Depression with programs such as the Works Progress Administration and Civilian Conservation Corps. Revisionists claim now that it was World War II, not FDR's "make-work" programs, that cured the economy. Some say it was both.

The newcomers in Washington venerate lower taxes, balanced budgets, and a sound dollar. "The stimulus has failed to stimulate anything but the deficit," declared Rep. Kevin McCarthy (R-Cal.), who helped draft the "Pledge to America," which calls on the 112th Congress to immediately cut $100 billion from the budget and stand firm against tax increases of any kind.

So much for President Obama's proposal last fall to create a $50 billion infrastructure "bank" to further stimulate surface transportation projects. Republican leaders are instead trying to claw back unspent funds from the original 2009 stimulus. That may prove futile because almost all the unspent money intended for public works has been technically obligated to specific projects and probably cannot be retrieved.

TRAIN WRECKS

One set of projects that are being cancelled—at least for the time being—involves proposals for intercity and high-speed passenger rail service between major Midwestern cities.

"That train is dead," exclaimed John Kasich, the Republican ex-Congressman, after being elected governor of Ohio in November and putting an immediate ki-bosh on plans for a $400 million Cleveland-Columbus-Cincinnati passenger rail corridor. Shortly thereafter Wisconsin's new Republican governor-elect, Scott Walker, confirmed that, per his campaign promise, plans were being halted for an $810 million Milwaukee-to-Madison line. That leg was central to the U.S. Depart-ment of Transportation's longer term designs for a Chicago-to-Minneapolis bullet train.

"This recommitment to the automobile is not well-timed," Rick Harnish, presi-dent of the nonprofit Midwest High Speed Rail Association, said in a statement critical of the cancellations. "With gasoline prices forecast to keep on rising," he adds, "it is urgent that we create other ways to get around."

No doubt governors Kasich and Walker were steeled in their resolve by New Jersey Gov. Chris Christie's decision, announced just weeks before the midterm elections, to cancel his state's participation in a new tunnel to carry passenger trains under the Hudson River and into Manhattan.

"I cannot place upon the citizens of New Jersey an open-ended letter of credit, and that's what this project represents," Christie said in scotching what was to be, at $9 billion, the largest public works project in the nation. New Jersey and New York each would have contributed a third of the tunnel's bid price, but the states would have been responsible for any cost overruns.

In Gov. Christie's rhetorical run-up much was made of Boston's experience with the Big Dig. That massive redo of America's oldest urban waterfront was initially estimated at $2.8 billion and ended up costing $14.6 billion, not including interest on construction bonds. Little wonder the project has become a "Remember the Alamo" cry among deficit hawks, regardless of any positive results that followed the project's completion. (See "Getting Around Gets Easier," January 2011.)

Advocates of the Hudson River tunnel, which would have significantly reduced commuting times for more than a half million riders, called Gov. Christie's ap-proach "shortsightedness." Sen. Frank Lautenberg (D-N.J.) called it "one of the biggest public policy blunders in New Jersey history."

A WAKE-UP CALL

Regardless of whether it was a blunder, it was certainly a wake-up call for plan-ners and public officials, a call that echoes through the halls of Congress. The old arguments for investing in public works, like the old cost-plus ways of building them, just aren't cutting it anymore. Not when governments at every level are hamstrung by unpaid bills and bonded indebtedness, by underfunded pensions and unaffordable entitlements.

"The building binge is over," says Steve Elkins, a city council member from Bloomington, Minnesota, and immediate past chair of the transportation and in-frastructure steering committee of the National League of Cities. "Now our mantra has got to be 'fix it first.' We're looking for low-cost, high-benefit projects—squeeze in an extra lane here, realign an interchange there. We took our stimulus money and did one-and-a-half projects worth of bottleneck relief."

Given what's happening in Washington, Elkins says, large hoped-for projects such as a light-rail line connecting the downtowns of the Twin Cities with their southwest suburbs "probably goes back on the shelf." Chances are that the same can now be said of big-ticket projects on TIPs lists all over the country.

David Sander, the National League of Cities' new infrastructure chair and a councilman in Rancho Cordova, California, sees this new fiscal circumstance as a sign that states and localities have entered a period of "pragmatism and devolution." His relatively young Sacramento suburb has made great use of developer contributions, not just to pay for adjacent streets and sewer hookups, but to fatten up a flexible central fund used for planned improvements all over town.

Sander worries, however, about the current attack against earmarks. He explained that strict federal allocation formulas, calibrated as they are by state and by category, don't anticipate special needs. It took a specific earmark, he says, to obtain federal dollars to jump-start a facelift of Rancho Cordova's old U.S. Route 50 commercial corridor, a once-thriving strip rendered lifeless by a paralleling interstate.

He also puts in a plug for stronger intergovernment cooperation. Only by involving two cities, two counties, and CALTRANS, he says, was Rancho Cordova able to win an extension and widening of White Rock Road into the foothills east of town. Next up: a possible toll road connecting U.S. 50 and California 99 south of town.

Toll roads, Sander concedes, are a hard political sell. Much of the voting public thinks expressways appear like manna from heaven. But user fees are an obvious alternative source of funding now that Washington is pulling back.

MISSING A BIG SALE

What galls many observers is the unfortunate timing of the federal pullback now that dollars spent on public works are buying so much more. Money is cheap because interest rates are at 50-year lows. Labor and most materials are cheap, what with the moribund state of the housing and office markets. Unemployment in the construction sector has been running about 17 percent, according to the government's Bureau of Labor Statistics. Still worried about cost overruns?

Consider this: For all the delays experienced by those not-so-shovel-ready stimulus projects, an early U.S. DOT audit determined that bids and contracts were coming in roughly 20 percent below original estimates.

"There's work that needs to be done," a rueful Lawrence Summers told *Newsweek.com* before his resignation as chairman of the National Economic Council was announced last fall. "There are people there to do it. It seems a crime for the two not to be brought together."

BULLET TRAIN EXCESS?

Then again, some think Washington's big reversal may prove a blessing in disguise if it provides a chance to rethink the nation's infrastructure priorities. Consider the momentum that had been building behind high-speed rail, not the 100-mph variety powered by existing diesel locomotion but the 220-mph high-tech bullets pioneered by Japan and France, and most recently built in China.

Some infrastructure advocates are wondering out loud about the cost-effectiveness of a technology that, in its purest form, requires dedicated rights-of-way (no sharing with freight trains), seamless tracks without grade crossings, and next-generation stations and terminals. Yet what seems a pipe dream to skeptics got real in a hurry when the Obama administration and a Democratic Congress included an initial $8 billion for high-speed engineering and feasibility work in the stimulus package launched in February 2009.

Then again, the administration's definition of "high speed" is trains that can be "reasonably expected to reach speeds of at least 110 miles-per-hour." So several proposed corridors are merely improvements on existing diesel locomotion—typically by elimination of freight conflicts—though others, like the Los Angeles-to-San Francisco route, call for full-bore 220-mph bullet trains.

There are 14 high-speed corridors in all, and in December U.S. Transportation Sec. Ray LaHood announced that the $1.2 billion spurned by Ohio and Wisconsin will be reallocated among them.

Of course, these are just startup grants. Total project cost for the LA–San Francisco line, for instance, has been estimated at $45 billion, although many think that a gross lowball number.

"We support a good intercity passenger rail system, one that is fast, efficient, and reliable," says Frank Beale, a bullet train skeptic and executive director of the Chicago area's Metropolis 2020 business leadership group. "But at what cost do you go twice as fast?"

"If I had access to that kind of money," Beale says, "I'd use it on intracity transit. It's a question of priorities. Let's first take care of what we have."

PLENTY TO FIX

And much of "what we have," Beale reminds us, is in pretty awful shape. The Chicago area's Regional Transportation Authority has identified a $17 billion "backlog" of needed repairs to its subways, buses, and commuter rail systems . . . and the "to do" list grows by about $300 million per year. Example: Chicago's not-so-rapid elevated transit lines traverse more than 500 street viaducts, many nearly a century old, several now requiring "temporary" wooden bracing.

Yet even rust-pocked Chicago gets distracted by fanciful plans. The centerpiece of the city's long-range downtown plan, a proposed West Loop Transportation

Center, would be a subterranean, multimodal, multilevel, rail-bus-pedway extravaganza that all but screams "Big Dig!"

But even if America were to limit its infrastructure appetite to the things we truly need, the immediate outlook is not promising. Not when a majority of the voting public equates public works with the politics of pork. Not when the new drive for fiscal austerity is allowed to push aside the very public investments needed to restore economic productivity.

There was, arguably, a small but significant cause for optimism in last November's election returns. Amy Liu, senior fellow and deputy director of the Brookings Metropolitan Policy Program, tracked 30 sizeable public works referendums around the country and found that three-quarters of them passed. When voters see an infrastructure improvement as immediately beneficial to their own community, she argues, they seem willing to tax themselves for it.

REBUILDING PUBLIC SUPPORT

How, though, do we build a popular constituency for a smarter electrical grid, or a freight rail overpass, or any number of out-of-sight, out-of-mind improvements that, to the untrained eye, seem lacking in here-and-now benefit?

One way is for planners to advocate for infrastructure whenever the opportunity presents. "Planners must act as key leaders in preparing our communities to face these new challenges, and help them become places of lasting value," urges the preamble to a report issued last fall by Rebuilding America, APA's National Infrastructure Investment Task Force.

Although it was released a month before last fall's political catharsis, the report also acknowledges that "we can no longer just engineer our way out of our infrastructure problems," and "we have no choice but to evolve away from the inefficient and unsustainable practices of the past."

And this, ultimately, may be the best take-away from last November's election. The game has changed—perhaps necessarily so given the new fiscal realities. How we respond may well determine whether our country—and our country's communities—recover or regress in the competitive future that awaits.

JOHN MCCARRON *is a Chicago-based urban affairs writer and contributing op-ed columnist for the* Chicago Tribune. *The opinions expressed in this article are his—with the exception of those quoted.*

RESOURCES

From APA

Rebuilding America, APA's report on infrastructure investment, is at www.planning.org/policy. See also "New House Rules Threaten Cuts in Transportation Programs," at http://blogs/planning.org/policy.

Tracking dollars

Estimates of infrastructure needs are detailed by the American Society of Civil Engineers at www.infrastructurereportcard.org.

Find documents from the president's commission on fiscal responsibility and reform at www.fiscalcommission.gov.

Track how federal recovery funds are being spent at www.recovery.gov.

Pro-infrastructure opinion and blogging are at www.infrastructurist.com.

Look Out Below!*

America's Infrastructure Is Crumbling

By Eric Kelderman
Stateline.org**, January 22, 2008

The numbers are staggering. More than one in four of America's nearly 600,000 bridges need significant repairs or are burdened with more traffic than they were designed to carry, according to the U.S. Department of Transportation.

A third of the country's major roadways are in substandard condition—a significant factor in a third of the more than 43,000 traffic fatalities each year, according to the Federal Highway Administration. Traffic jams waste 4 billion hours of commuters' time and nearly 3 billion gallons of gasoline a year, the Texas Transportation Institute calculates.

Dams, too, are at risk. The number of dams that could fail has grown 134% since 1999 to 3,346, and more than 1,300 of those are "high-hazard," meaning their collapse would threaten lives, the Association of State Dam Safety Officials (ASDSO) found. More than a third of dam failures or near failures since 1874 have happened in the last decade.

Underground, aging and inadequate sewer systems spill an estimated 1.26 trillion gallons of untreated sewage every year, resulting in an estimated $50.6 billion in cleanup costs, according to the U.S. Environmental Protection Agency.

"Much of America is held together by Scotch tape, bailing wire and prayers," said Donald F. Kettl, director of the Fels Institute of Government at the University of Pennsylvania.

Fixing these problems and others threatening the nation's critical infrastructure would cost $1.6 trillion—more than half of the annual federal budget, the Ameri-

---

* News organizations and other websites and publications are welcome to re-publish Stateline news stories and graphics unchanged, with attribution and a disclosure statement that reads as follows: "Stateline.org is a nonpartisan, nonprofit news service of the Pew Center on the States that reports and analyzes trends in state policy."
** Stateline.org was a project of the Pew Research Center from 2004 to 2008. As of July 1, 2008, it is a project of the Pew Center on the States.

can Society of Civil Engineers (ASCE) estimates. And that doesn't include what it will cost for new capacity to serve a growing population.

Recognizing the importance of structures so integral to U.S. commerce and Americans' well-being and safety, local, state and federal governments already are budgeting nearly two-thirds of the $1.6 trillion needed for infrastructure work. The problem is they raid many of those funds for other purposes, ASCE says.

Coming up with new money to fill the funding gap has become a political nightmare, with politicians and the public trying to avoid anything that looks like a higher tax.

"We have convinced ourselves that infrastructure is free, that someone else should be paying or that we have paid our share," said Mike Pagano, an urban planning expert at the University of Illinois at Chicago.

Infrastructure is the four-syllable jawbreaker that governments use to describe the concrete, stone, steel, wires and wood that Americans rely on every day but barely notice until something goes awry. Broadly speaking, it includes airports, the electrical energy grid, hazardous and solidwaste storage sites, navigable inland waterways, public parks, schools and even the security to protect all of those structures.

While the federal government bears the broadest responsibility to keep America's gears turning, state and local governments are accountable for supplying more than half of the money and all of the manpower to build and maintain the country's vast ground transportation network. States also have regulatory oversight of 85 percent of dams and help fund drinking-water and wastewater systems. Federal and state officials share the blame for shortfalls in America's maintenance budget. Congress hasn't raised the federal gasoline tax of 18.4 cents per gallon—which pays for about 45 percent of all road construction—since 1993, nor have many state leaders been willing to charge drivers more at the pump to pay for local road repairs.

The association of state dam officials contends that most state dam safety programs are underfunded, understaffed and often don't have adequate authority to regulate safety standards or emergency plans. Likewise, the federal dam safety program, which helps pay for the upkeep of structures, never has been fully funded by Congress.

The EPA estimates that the nation is falling short on water infrastructure by $22 billion annually. The federal Clean Water State Revolving Fund, which makes low-interest loans to clean up or protect water supplies, has shrunk from more than $3 billion in 1990 to roughly $1 billion in 2007.

The consequences of skimping can be dire:

- On Aug. 1, 2007, the Interstate 35 bridge in downtown Minneapolis collapsed into the Mississippi River, killing 13 people and injuring at least 80. Losing the state's most heavily traveled bridge is costing an estimated $400,000 daily in extra commuting time and gasoline, said Brian McClung, a spokesman for Minnesota Gov. Tim Pawlenty (R).

(A report issued Jan. 15 by the National Transportation Safety Board blamed the bridge collapse on inadequate steel "gussett" plates that hold the structure's angled beams together.)

- Steam pipe explosions in Midtown Manhattan last summer killed one person, injured dozens and disrupted businesses.
- In March 2006, the 116-year-old Kaloko Reservoir Dam in Hawaii collapsed after heavy rains, killing seven people and causing nearly $15 million in damage.
- In August 2005, after Hurricane Katrina, levees holding back Lake Pontchartrain gave way, flooding major parts of New Orleans. The storm and flooding are blamed for more than a thousand deaths and more than $100 billion in damage.
- In May 2002, the Interstate 40 bridge near Webbers Falls, Okla., collapsed into the Arkansas River, killing 14 people.

Despite urgent calls to prevent more tragedy from failed infrastructure, politicians and voters have signaled they are gun-shy of new taxes. After the collapse of the Minneapolis bridge, Minnesota politicians failed to agree to a statewide transportation package, putting off to the 2008 legislative session more debate over a proposed 5-cent hike in the state's gasoline tax. Gov. Tim Pawlenty (R) twice vetoed gas-tax hikes before the bridge fell.

Read the full report at stateline.org.

American Collapse[*]

By Sarah Williams Goldhagen
The New Republic, August 27, 2007

I.

Within fourteen days of each other, two rush-hour calamities: a bridge collapse and a steam-pipe explosion. In Minneapolis, a forty-year-old bridge along highway I-35W suddenly dropped sixty feet into the Mississippi River, killing at least five people and injuring approximately one hundred more. The federal government had deemed the bridge structurally deficient in 1990, which the Minnesota Department of Transportation acknowledged in separate reports issued in 2005, 2006, and 2007, after inspecting the bridge. In seventeen years, federal and state agencies repeatedly reported significant problems with the fallen bridge and yet no meaningful repairs were made. In midtown Manhattan, meanwhile, an eighty-three-year-old asbestos-wrapped steam pipe exploded, killing one person and injuring dozens more. That pipe is owned by the private utility company Con Edison, whose crews had inspected it earlier the same day and deemed it safe.

Headlines screeching news of these two horrifying events have replaced, temporarily, the usual newspaper rhythm of weekly incantations announcing this or that city's plans for adorning itself with a new stadium, public park, or luminescent museum—announcements that often serve to distract the public's attention from the silent scourge afflicting this country's viscera. One pipe explosion and one bridge collapse just might be enough to rouse the public to the news that America's metropolitan regions are in serious trouble. Bridges, utilities, and flood-prevention systems, whether publicly or privately owned, are grossly neglected. Suburbs are sprawling like unchecked chickweed. Cars are stuck in ever-mounting hours of traffic. Cities are bleeding people. School buildings are overpopulated and crumbling. Waters are polluted. Shipping ports are decrepit.

Typically, these matters are discussed piecemeal, as discrete problems beleaguering this or that city or suburb (Minneapolis, Manhattan) or this or that infrastructural element (bridges, utilities). Politicians, voters, and the professional stewards of our built environment—city planners, architects, landscape architects, urban designers, civil engineers—could let inertia run its course and continue to approach such problems, conceptually and politically, in this balkanized manner. If they do, the American media will continue to feast upon these tales of mismanagement and woe, and only marginal, localized improvements will occur. But these problems are not discrete or local. They are part of a larger single phenomenon: America's failure to manage the physical plant of its urban settlements, to maintain what it should while designing, funding, and building desperately needed new facilities.

This country's negligence of its physical plant, its extent and its far-reaching implications, becomes especially evident if we conceive of our urban settlements differently. Dispense with the categories of city, suburb, and exurb; dispense absolutely with the dichotomy of city versus suburb. Instead, consider the city-suburb-exurb nexus as the interwoven entity that it now is: a metropolitan region. This was a principal point of last year's Venice Architecture Biennale, "Cities, Architecture, and Society," a multimedia extravaganza about metropolitan regions around the globe, filled with colored maps and three-dimensional bar-graphs projecting global shifts from 2000 to 2050 in population, demography, literacy rates, and economic growth; exhibiting high-quality photographs documenting the current state of metropolitan regions; and displaying architectural, urban design, and urban planning projects addressing those changes. The related exhibition "Global Cities," curated by Richard Burdett, a professor of architecture and urbanism at the London School of Economics, opened in June in the Turbine Hall of the Tate Gallery in London. In the exhibition's eight-hundred-page catalogue, *Cities: People, Society, Architecture*, Burdett predicts that by 2050, 75 percent of the world's peoples will live in metropolitan regions. More than 80 percent of Americans already do.

Altering our view of the United States to that of a country composed of a collection of metropolitan regions helpfully relegates the local and particularistic qualities of these problems to the background and highlights their commonality instead. And what is that commonality? As Frances Halsband of Kliment & Halsband, a distinguished architect on the Architectural Review Board of the Federal Reserve Bank and the Architecture Advisory Board of the U.S. State Department, puts it in the Architectural League's recent exhibition "New New York: Fast Forward": "Infrastructure, infrastructure, infrastructure."

Infrastructure is one crucial point at which politics and architecture merge. A country's physical plant should be front and center in the policy agendas of its public officials, and it should be front and center in the intellectual and professional agendas of the professional stewards of its built environment. For many reasons, this is not the case in the United States. The very notion of infrastructure is financially, physically, and conceptually unwieldy: it encompasses highways, street systems, sidewalks, schools, bicycle paths, mass transportation systems, communication systems, utilities, sewage and water treatment plants, public parks, and complexes of institutional facilities serving civic, cultural, or leisure activities. But infrastructure should be defined not by what it looks like, and not by who designs it or who pays for it, and not by who builds it or actually uses it. It should be defined by whom it is meant to serve. For all its seemingly disparate parts, infrastructure comprises those elements in a metropolitan region's physical landscape that are meant to serve the public—or rather, the sometimes competing, sometimes overlapping, and sometimes wholly discontinuous publics that populate today's American metropolitan areas and are critical to the growth of our country.

A quick survey of the infrastructural elements of the United States' metropolitan regions suggests that a few might be said to be doing tolerably well. The rest, which means those in most of the country, are in horrendous shape. Large swaths of our infrastructure—not just one bridge in Minneapolis, or even a bunch of bridges across the country, or a bunch of asbestos-wrapped steam pipes coursing under our cities' streets—have aged to the point of gross deterioration. To sense the magnitude of the problem, one need look no further than the sobering "Infrastructure Report Card" on the United States, which is published every few years by the country's leading professional organization in the field, the American Society of Civil Engineers. On the ASCE's most recent Report Card, from 2005, not a single one of fifteen categories received a grade higher than a C. Ten of the fifteen categories—including drinking water, waste-water management, navigable waterways, transit, and schools—received scores in the D range.

These grades are not subjective evaluations. Twenty-four of the nation's leading civil engineers analyzed hundreds of reports and studies and surveyed more than two thousand of their colleagues before assigning them. The grades are abundantly substantiated with statistics, the results of research studies, and estimates of the dollar figures necessary for remediation. The ASCE's report on the conditions in most American metropolitan regions reveals a shocking national indifference to the maintenance, upgrading, and creative re-conceptualization of our own infrastructure. The ASCE Report Cards from 2003, 2001, and 1998 do not look better.

But never mind grades. Consider facts and numbers. The explosion in Manhattan damaged a number of feeder cables for the city's electric system, closing more than one hundred businesses with an estimated loss of income of $10 million to $30 million, and leaving a crater nearly forty feet in diameter that will disrupt traffic for weeks in one of the city's busiest neighborhoods. The estimated figures for cleaning up the mess in Minneapolis and building a new bridge approach $400 million. And these figures are only for the two most recent catastrophes. More than 160,000 of the nation's nearly 600,000 bridges are structurally deficient or functionally obsolete, with I-35W having been one of them. The National Priorities List currently includes 1,237 toxic waste sites, and the Environmental Protection Agency has identified more than 350,000 contaminated sites that need to be cleaned up within the next three decades. Americans spend 3.5 billion hours annually stuck in traffic. In 2000, the National Education Association reported that more than $268 billion was needed to bring the nation's school facilities into good overall condition.

A quick survey of some of the ASCE's monetary estimates betrays this appalling indifference and the catastrophe that it portends. America faces an annual shortfall of $11 billion to replace aging facilities in order to comply with safe drinking water regulations, and yet as of 2005, federal funding for drinking water remained at less than 10 percent of this total. The National Park Service estimates (and probably underestimates) a maintenance backlog of $6.1 billion. In January dozens of beaches in California were closed when heavy rains caused overflow that dumped millions of gallons of raw sewage into the waters. The number of unsafe dams rose by 33 percent between 1998 and 2005. The number of non-federally-owned dams identified as unsafe is increasing at a faster rate than those being repaired. For all the non-federally-owned dams that pose a direct risk to human life if they should fail, the ASCE estimates that $10.1 billion is needed over the next twelve years to make them safe. In sum, the ASCE estimates that the country would need to spend $1.6 trillion in the next five years to bring the country's infrastructure to an acceptable minimum standard.

And that's just getting what we already have back into shape. What about meeting the myriad needs of the reconfiguring world of the twenty-first century? The United States has a rapidly growing population, with certain regions, especially in the Sun Belt, requiring vast new infrastructure. Everywhere there continue to be shifts in population farther and farther from the core cities, making metropolitan regions all that much more far-flung and posing new infrastructural problems and demands. At the same time, we must cope with the phenomenon of "Shrinking Cities," the title of an exhibition at the KW Institute for Contemporary Art in Berlin in 2004 and of the indispensable three-volume catalogue that accompanied it. When a city's population significantly declines—which has happened in cities as different as New Orleans, the District of Columbia, Cleveland, Pittsburgh, and St. Louis, each of which has lost 30 to 50 percent of its inhabitants in the past half-century—large abandoned swaths of land are left behind. Do we leave de-populated cities to fend for themselves, with their boarded-up buildings deteriorating and HAZARD signs posted on building fronts, creating self-fulfilling prophecies of cities' deaths foretold? Do we neglect the opportunity to develop creative new uses for our nation's waterfronts, many of which are no longer adequate to serve as railway transit or shipping ports, and which currently sit as wastelands in sites as prominent as the west side of Manhattan? Do we ignore the thousands of brown-fields—which the Environmental Protection Agency defines as "real property, the expansion, redevelopment, or reuse of which may be complicated by the presence or potential presence of a hazardous substance, pollutant, or contaminant"—that defile our purple mountains' majesty? Do we just keep building highways and airports, and more highways and airports, when in truth we must begin to reconfigure America's physical landscape to better accommodate its citizens' changing demographic profile, social habits, and economic needs?

How did the United States devolve into this perilous condition? Why has the world's greatest economic and military power neglected what is arguably its most important asset after its people (also much neglected): the infrastructure that literally constitutes its cities and metropolitan regions? How are people choosing, or being shut out of choosing, the physical configuration of their metropolitan regions, from domicile to parkland to sewage-treatment plant? The answers to these questions must come from politicians at the local, the state, and especially the federal levels; and from city planners and civil engineers; and from urban designers, architects, and landscape architects. Unless the problem is viewed with a wide-angle lens that reveals its many mutually reinforcing dimensions, it cannot be properly analyzed.

Visit metropolitan regions in many of the rapidly developing countries in Asia, and the gross inferiority of America's physical infrastructure is immediately apparent. Highways and roads in those countries are not pitted moonscapes. Public transportation, from trains to trolleys to buses, is plentiful, in good repair, and punctual. Public structures of all kinds—from governmental and civic buildings to public parks and urban plazas to "streetscape" elements such as pedestrian bridges and roadway lighting fixtures—are of immensely higher design quality and in immensely better shape. At every turn, a citizen moving through the built environment of these countries sees and uses physically embodied signs communicating to them that in their society, public life matters.

But forget, if you wish, the vast infrastructural building taking place across economically exploding Asia and the Middle East. Look no farther than Europe or Canada, areas in what used to be called the industrialized world, where metropolitan regions are facing the same problems of demographic shifts, higher labor costs, and aging infrastructure that we face in the United States. Again and again we find examples of metropolitan regions that have successfully risen to these challenges. Two of the most extraordinary recent success stories are Barcelona and Vancouver. Barcelona, in preparation for the Summer Olympics in 1992, remade itself by hiring world-class city planners, urban designers, architects, and civil engineers to rehabilitate its aged urban center, construct an Olympic Village that has become a new settlement in its metropolitan region, construct a new ring road, and reclaim its waterfront by transforming miles upon miles of beaches into parkland with bicycle paths, marinas, swimming areas, urban plazas, and high-quality residential fabric—all accessible by public transportation. And Vancouver's physical remaking in the past two decades has been so remarkable that it has become a phenomenon, a brand: the "Vancouver Miracle," a city that, twenty years ago, was an emptied-out downtown littered with disused industrial lots and is today a lively, high-density, twenty-four-hour city filled with attractively designed high-rise residential and office towers, well-preserved historic buildings, plentiful public parks, and vibrant cultural institutions. Vancouver is currently the fastest-growing residential downtown in North America.

Infrastructure is an amorphous topic. Building it and maintaining it is staggeringly expensive. In the eyes of many voters, elected officials, and professional stewards of the built environment, maintaining it, trying to visualize what it should be or could be, is just plain boring. Re-orienting ourselves toward our infrastructure and bringing it up to date requires clearing away multiple political, social, and conceptual obstacles. As this past spring's exhibitions on Robert Moses in New York City made clear about the politics of American infrastructure, the nature of the American democratic system discourages the long-term planning, the commitment to public investment, and the vertical coordination of federal, state, and local initiatives that most advanced industrial countries enjoy, which are fundamental to the establishment, maintenance, and upgrading of the existing infrastructure, not to mention the initiation of new projects.

American citizens know that they and their children need safe bridges, clean drinking water, public school buildings that do not force classes to convene in hallways or leak or cause sick building syndrome, and clean and healthy parks so that playing children are not forced onto city streets or toxic waste sites. Yet voters are exquisitely prone to sticker shock, and large-scale infrastructural projects—a new mass-transit line, a new park or waterway system, a bridge—cost an extraordinary amount of money. Indeed, the projected dollar amounts themselves are often enough to cause the political equivalent of what psychologists of trauma call dissociation. (Did you say $14.6 *billion* for Boston's Central Artery project?) Even smaller-scale projects—a new public school, a waterfront park (such as Pittsburgh's $22 million Allegheny Riverfront Park, or the recently completed $85 million Olympic Sculpture Park in Seattle), an urban public plaza (such as the $650 million retooling of Lincoln Center in New York City to make it more user-friendly and accessible to the public)—seem fantastically expensive to anyone who has no experience of planning and managing costs for any major construction project—in other words, to nearly everyone, including nearly every one of our voting citizens.

When a proposed tax or budget item can be specifically linked to the officials whom voters elect to their own state or local offices, the political disincentives to address a metropolitan region's infrastructural needs are enormous. Funding maintenance for existing infrastructure is doubly challenging because there is not even the political reward of a shining new public amenity. If Madame Governor proposes a good or necessary infrastructure project, should she keep trying to push the project through when the bidding process is complete and her constituents balk at its projected cost? Should she do so if the ultimate cost is her job? Should she do so with the knowledge that, at any point along the way, from conception to design to construction, determined and vocal detractors can point to the mounting dollar signs, and portray that project as a sinkhole for taxpayer money, and derail it?

Even in the heyday of American infrastructure-building, from 1930 to 1970, it took an imperious wheeler-dealer such as Robert Moses to take maximal advantage of the funds that the federal government was making available to American cities: owing to Moses, for example, New York City received more than twice the Title I funds for slum clearance of any other city in the country. Robert Caro, Moses's biographer, who was simultaneously fascinated and revolted by Moses's labyrinthine anti-democratic conception of his calling, conceded in *The Power Broker* that "the problem of constructing large-scale public works in a crowded urban setting, where such works impinge upon the lives of or displace thousands of voters, is one that democracy has not yet solved." American democracy, that is.

The neglect of infrastructure has dramatically worsened since the 1970s, for two reasons. First, the country has undergone a structural transformation from city-suburb-exurb-farmland, a constellation that does not necessarily conflict with the tripartite local-state-federal structure of our government, into metropolitan regions, a constellation that does conflict with that structure. We are stuck with the existing political, legal, and institutional structures of states (usually bigger than metropolitan areas) and municipalities (smaller and self-interested) through which almost everything must be organized and funneled. Neither is the right kind of entity for managing a metropolitan region, but together they inevitably organize our thinking and, more important, our policy planning, which turns out to be too unfocused (in the case of states) or too hyper-focused (in the case of municipalities).

Second, the federal government has increasingly fobbed off the responsibility for maintaining and upgrading the country's infrastructure onto state and local governments, which bear the legal obligation to attend to their regions' infrastructure but are systemically constituted in such a manner that only rarely can they command the financial resources adequate to accomplish the necessary tasks. A Congressional Budget Office study reveals that an immediate and steady decline in federal spending on infrastructure as a percentage of GDP began with Reagan's first budget and continues straight downhill, with the average annual amount in 1982–1998 being 29 percent less than in 1965–1981. While in the earlier period the average was .93 percent of GDP, in the last year of the CBO's study the downward trend had brought spending all the way down to .57 percent of GDP—a drop of almost 40 percent. Federal spending on infrastructure as a percentage of all federal spending between these two periods, which include Republican and Democratic administrations, declined even more, by 33 percent. And states and municipalities have not picked up the slack: state and local spending throughout these years has hovered around a little less than 2 percent of GDP. Max Sawicky of the Economic Policy Institute uses somewhat different measurements, though his figures for public net investment in physical capital come up to 2006: he estimates that for the ten years between 1959 and 1969, public net investment in physical capital was about 2.6 percent of GDP; but for the last ten years it has been about 1.1 percent.

Without a large federal role, infrastructural needs cannot be effectively addressed, in part because the American political system has always made it difficult for state and local governments to do so, and in part because metropolitan regions cut across municipal and sometimes even state lines. This is one of those tasks that only the federal government can accomplish. A dramatic example of why state and local governments cannot go it alone on infrastructure building comes from New Mexico, which has recently completed a new four-lane highway: when that new highway reaches the Colorado state border, it dies into two lanes, because the state of Colorado has not coughed up the funds to continue it. Owing to the disincentives to address the country's infrastructure that are built into the American democratic system—and to the rightward shove of our governance in the last twenty years, which has militated against even discussing, let alone addressing, large-scale public needs—the political obstacles to taking America's infrastructural problems seriously are enormous. Shouldn't we think about our country's physical plant in terms not that different from our legal and regulatory systems in general—as a necessary foundation for the social and economic health and growth of this country, requiring substantial federal leadership and funding? We need a national infrastructure for infrastructure.

II.

Those are not the only reasons for this American crisis. There is also a cultural obstacle: the public's lack of faith in expertise when it comes to the planning and the construction of our physical environment. This is truly bizarre. When you have legal needs, you go to a person trained to deal with them, to a lawyer. For medical needs, to a doctor. For electrical needs, to an electrician; for investment needs, to a financial expert. But for the needs of infrastructure and urban planning and design, the politicians and the public are leery if not downright dismissive of the experts trained to deal with such problems: city planners, urban designers, architects, landscape architects, and civil engineers.

There are many reasons for the low esteem in which the public currently holds these experts in the built environment. Some are justified, others are not. It began with the widely publicized failures of the federally funded slum clearance and urban renewal programs of the 1960s, which nurtured a crude morality tale of the consequences of governmental intervention in the country's physical plant, a David-and-Goliath melodrama played out in the standoff between a feisty little lady named Jane Jacobs and an outsize predatory "expert" named Robert Moses. An overly simplistic and misleading fable was born, according to which government should keep its grubby hands off our cities. Nearly half a century later, this legend continues to dominate public thinking about how our country's infrastructure should be managed.

Today's city planners are seen as clueless and well-meaning bureaucrats at best, and as anti-democratic elitists at worst. Architects, landscape architects, and urban designers (including the many who do not merit the slander) are depicted as divas who care more about fancy forms than about the people who live in their buildings or the clients who build them. Expertise in the built environment is often held in public ridicule. As a result, folk wisdom has it that it is up to the public-spirited citizen—the community board activist, the local environmental review agency, the historic preservation commission—to stop them: thus unwittingly validating a salient quotation that was prominently displayed in one of the Moses exhibitions. "The critics," he once said, "build nothing."

Sadly, the public's mistrust of the experts who should be advising politicians on how to address the multifarious problems of the American infrastructure is not wholly misplaced. To be sure, there are many talented public officials, city planners, urban designers, and architects who are committed to working in the interest of the public good. Yet there are also many professionals who have resigned themselves to working within the ever-narrowing constraints that the public assigns to them of designing mainly signature projects. The most self-damning example of this was the architecture and design community's befuddled and largely negative reaction to Burdett's call-to-arms "Global Cities" exhibition. The maps, the bar graphs, and the statistics confounded them. The exhibition was roundly condemned as "boring." Such reactions are all too common. As a result, American architects sometimes run the risk of appearing to be little more than glorified shoe designers.

When evaluating the professional practice of American designers, we need to consider their substantial contribution to the current infrastructural morass. American design professionals have failed to conceptualize, or to communicate effectively, theoretical and practical visions of how the American infrastructural landscape might be reconfigured to address the challenges of the new century. In the architecture and urban design community, for example, only two visions— both glaringly inadequate—currently dominate the discourse about the future of urban life: one propounded by the Congress for New Urbanism, founded largely by Andreas Duany and Elizabeth Plater-Zyberk, and another put forth by the Office for Metropolitan Architecture, founded by Rem Koolhaas.

The CNU, inspired by the ideas in Jacobs's *The Death and Life of American Cities*, promotes a more or less sensible set of principles for urban development that are now commonly accepted: zoning should nurture a mix of commercial, retail, and residential uses; density is preferable to sprawl; nodal communities are preferable to bedroom suburbs; infrastructural planning should prioritize mass transit over the automobile; and so on. Yet the problem with the New Urbanism (with the exception of some excellent work being done in California by Peter Calthorpe and others) is that for all Duany's rhetorical bluster and for all the CNU's talk of heterogeneity, actually executed New Urbanist plans tend to be excessively legislated, code-bound, and architecturally reactionary. The two most famous New Urbanist projects are unbearably homogeneous, socially and architecturally: Seaside, in Florida, the set for the movie *The Truman Show*; and Celebration, also in Florida, built by the Disney Corporation and as fanatically controlled as any of that company's other products.

The only serious high-profile competitor to the CNU as a vision for the twenty-first century has been trumpeted by Koolhaas and his many far-flung acolytes; but Koolhaas's "everything is funky, let it be" attitude amounts to little more than developer-friendly urbanism and is no more satisfactory. I know of only one promising new vision on the horizon, from James Corner of Field Operations in New York City. Field Operations is committed to an interdisciplinary approach to the infrastructural, ecological, and social problems facing metropolitan regions, and it draws from landscape architecture, architecture, urban design, and civil engineering in celebrated (but as yet unexecuted) projects, including prize-winning designs for the University of Puerto Rico and for the 2,200-acre Fresh Kills landfill in Staten Island.

So what, if anything, is being done for the infrastructure of our metropolitan regions today? Answering this question properly requires discipline. One must keep one's eye on the concrete. One must consider—and, when appropriate, applaud—well-meaning, carefully conceived, often highly publicized *plans*, such as Mayor Michael Bloomberg's 2030 plan for New York City, and Mayor Richard Daley's Central Area Plan for Chicago, which includes an initiative to make the Windy City "the greenest in America." And then one must turn away from what is on paper to examine what is taking place on the ground. And what is taking place? Well, not much, as the two recent catastrophes and the figures of neglect, decrepitude, and financial shortfalls tell us. And much of what is being done is pernicious.

Since city and even state governments can rarely undertake such large-scale initiatives on their own, the maintenance and the upgrading of our metropolitan regions' infrastructures is falling ever more into private hands. *Business Week* recently reported that dollar-starved city and state governments are selling off parts of their public infrastructures to the highest private bidder. Morgan Stanley, Merrill Lynch, Goldman Sachs, and other firms have started negotiating the purchase of, or have already bought up, a myriad of what once were and should properly be public facilities: roadways (pieces of the Pennsylvania Turnpike, the Colorado Northwest Parkway, the Chicago Skyway, and the Indiana Toll Road have already been sold), water systems (the city of Atlanta), and so on. The sales of the Tappan Zee Bridge, Chicago's Midway Airport, and Philadelphia International Airport have been explored.

In isolated instances, the privatization of the infrastructure of our metropolitan regions can have happy results. One example is the recently completed Olympic Sculpture Park in Seattle, a project designed by Weiss/Manfredi of New York that truly does serve the public and successfully manages many of the issues faced by many American metropolitan regions. The Olympic Sculpture Park reconnects downtown Seattle to its waterfront with V-shaped geometries that create a meandering, multilevel path beginning at the city's edge, straddling the linear gashes of a superhighway and active railroad tracks, and leading to a newly rehabilitated waterfront with a well-designed public esplanade. But the Olympic Sculpture Park, though a public amenity, was funded almost entirely through private philanthropic donations raised by the Seattle Art Museum, which abuts the site. It can hardly serve as a reproducible model for other cities and states looking for ways to manage the infrastructure of their downtowns or metropolitan regions.

More typical are the so-called public-private partnerships to which desperate cities across the country have been increasingly turning since the 1970s. One such project is Atlantic Yards in Brooklyn. Headlines on this project have predictably run this way: Brooklyn teams up with the developer Bruce Ratner to create a major new project that will contain private residences (which will profit the developer) and new public amenities. This project will jump-start the economic revitalization of downtown Brooklyn. Then a few years go by. New headlines: New York City has granted Atlantic Yards developer Bruce Ratner X, Y, and Z additional zoning, or land use, or other concessions. These concessions threaten to severely limit public access to the project. Yet if the concessions are not granted, the developer may pull out, to the economic detriment of the city. If Atlantic Yards is built as it is now envisioned, its public spaces are likely to suffer the same fate as those "privately owned public spaces" in Manhattan that Jerold Kayden meticulously documented in his book *Privately Owned Public Space: The New York City Experience*. Kayden studied the "public spaces" built by private developers from the 1960s onward in exchange for various land and air rights bonuses. He discovered that, absent vigorous oversight from municipal authorities, many of those spaces had been locked up, hidden away with plantings or other design elements, or so ill-maintained that one would have to completely redefine the meaning of "public" for them to qualify.

The privatization of the country's infrastructure is the current trend. In the short term it spells social disaster, and in the long term it spells economic disaster. The short term: will the private companies buying up our highways or airports be committed to maintaining free access and mobility to financially disadvantaged customers? Not likely. Will they even pretend to maintain—much less to upgrade—the infrastructure they own if it becomes financially more advantageous for them to put their resources toward some other enterprise instead? Not likely, as Con Ed's engineers inadvertently demonstrated with their inspection of the steam pipe the very morning of the day it exploded. Will private companies devise plans that will accommodate large social changes when they must be answerable to shareholders focused on annual reports? Not likely. And the long term: the piecemeal privatization of the nation's infrastructure will not, almost by definition, be effectively coordinated to create the continuous and continuously functional system on which this country's economic growth depends.

Even strict free marketeers should take pause at the less immediate but certainly inevitable economic consequences of this trend. Here is Adam Smith in *The Wealth of Nations*: it is the "duty of the sovereign or commonwealth" to erect and to maintain "public institutions and those public works, which, though they may be in the highest degree advantageous to a great society, are, however, of such a nature that the profit could never repay the expense to any individual or small number of individuals, and which it therefore cannot be expected that any individual or small number of individuals should erect or maintain." Infrastructure is the classic public good that the free market does not and cannot provide. On the scale that is necessary, only the federal government can make the difference.

Other advanced industrial countries (Canada, Spain, Germany, Japan, Korea—the list goes on) point the way to very successful and economically more viable paths toward addressing the infrastructural crisis. Policymakers at all levels of government and this country's design professionals must begin to accept that we are now a country of metropolitan regions. Then they must work in concert for change. Government officials at all levels, especially at the federal level, must take it upon themselves to address the maintenance, upgrading, and re-conceptualization of the nation's infrastructure. We are well beyond the point of needing a congressional commission to assess the state of the nation's infrastructure (though any action is better than none). We need a National Infrastructure Agency to plan, provide expertise for, fund, and coordinate long-term initiatives for infrastructural maintenance and improvement. At the very least Congress should establish a federal line-item capital budget, as most other developed countries have, which would help to reduce the perpetual scanting of long-term budgetary needs in favor of short-term ones. State officials should demand federal assistance to address their infrastructural needs. Municipal officials should find the legal and political mechanisms (such as metropolitan district authorities of broad scope) that would allow them to work in concert with, rather than in competition with, their counterparts in neighboring communities. Leading design professionals should refuse to be merely producers of high-end cultural icons and luxury housing, and make themselves relevant to every part of the infrastructure challenge, and work to reinstate the public's trust in their authority.

When infrastructural needs large and small have been addressed in the past twenty years, it has been because leaders decided to take action, and then refused to sleep or eat until their pet project was completed, or close to completion. How did Millennium Park in Chicago happen? Daley decided that it should happen and then persuaded people that he was right. How did Boston's Central Artery project happen? Fred Salvucci, then transportation secretary of Massachusetts, lived and breathed that project for nearly two decades, persuading local government officials, state authorities, federal representatives, and community boards that he was right. Why has the pathetically ill-conceived design for the twenty-seven-acre swath of reclaimed land created by burying Boston's Central Artery (I-93), ironically named the Rose Fitzgerald Kennedy Greenway, been hung out to dry? Because nobody stepped into Salvucci's shoes.

Surely ordinary Americans can recognize this crisis: they drive on it, cursing the traffic, every day. The ribbon of concrete and steel washing into the Mississippi River; the chasm on West 41st Street in New York (or the larger one—very different in origin, to be sure—that remains at New York City's Ground Zero); the boarded-up acres of urban disaster in New Orleans that Hurricane Katrina left behind; the billions of gallons of raw sewage released into our waterways every year; the stupendously mediocre Rose Kennedy Greenway; the bridges and the highways closed because of some spectacular failure or the need for emergency repairs—all these disasters are only the most publicly visible evidence of what happens, or does not happen, when policymakers and design professionals fail to effectively use their power for the oldest and best purpose of all, America's public good.

SARAH WILLIAMS GOLDHAGEN *is the architecture critic for* The New Republic.

Infrastructure Investment and U.S. Competitiveness[*]

By Jonathan Masters
Council on Foreign Relations, April 5, 2011

Most experts agree the United States must address the nation's aging network of roads, bridges, airports, railways, power grids, water systems, and other public works to maintain its global economic competitiveness. In 2010, President Barack Obama proposed a national infrastructure bank that would leverage public and private capital to fund improvements, and in April 2011 a bipartisan coalition of senators put forward a similar concept.

Four experts discuss how the United States can best move forward on infrastructure development. Robert Puentes of the Brookings Institution suggests focusing on increasing exports, low-carbon technology, innovation, and opportunity. Renowned financier Felix Rohatyn endorses the concept of a federally owned but independently operated national infrastructure bank that would provide a "guidance-system" for federal dollars. Infrastructure policy authority Richard Little argues that adequate revenue streams are the "first step in addressing this problem," stressing "revenue-based models" as essential. Deputy Mayor of New York City Stephen Goldsmith says that the "most promising ideas" in this policy area involve public-private partnerships.

Robert Puentes, Senior Fellow, Brookings Institution

Infrastructure is central to U.S. prosperity and global competitiveness. It matters because state-of-the-art transportation, telecommunications, and energy networks—the connective tissue of the nation—are critical to moving goods, ideas, and workers quickly and efficiently and providing a safe, secure, and competitive climate for business operations.

But for too long, the nation's infrastructure policies have been kept separate and apart from the larger conversation about the U.S. economy. The benefits of infrastructure are frequently framed around short-term goals about job creation. While the focus on employment growth is certainly understandable, it is not the best way to target and deploy infrastructure dollars. And it means so-called "shovel ready projects" are all we can do while long-term investments in the smart grid, high-speed rail, and modern ports are stuck at the starting gate.

We often fail to make infrastructure investments in an economy-enhancing way. This is why the proposal for a national infrastructure bank is so important.

So in addition to the focus on job growth in the short term, we need to rebalance the American economy for the long term on several key elements: higher exports, to take advantage of rising global demand; low-carbon technology, to lead the clean-energy revolution; innovation, to spur growth through ideas and their deployment; and greater opportunity, to reverse the troubling, decades-long rise in inequality. Infrastructure is fundamental to each of those elements.

Yet while we know America's infrastructure needs are substantial, we have not been able to pull together the resources to make the requisite investments. And when we do, we often fail to make infrastructure investments in an economy-enhancing way. This is why the proposal for a national infrastructure bank is so important. If designed and implemented appropriately, it would be a targeted mechanism to deal with critical new investments on a merit basis, while adhering to market forces and leveraging the private capital we know is ready to invest here in the United States.

Building the next economy will require deliberate and purposeful action, across all levels of government, in collaboration with the private and nonprofit sectors. Infrastructure is a big piece of that.

Felix G. Rohatyn, Special Advisor to the Chairman and CEO, Lazard Freres and Co. LLC

While America's economic competitors and partners around the world make massive investments in public infrastructure, our nation's roads and bridges, schools and hospitals, airports and railways, ports and dams, waterlines, and air-control systems are rapidly and dangerously deteriorating.

China, India, and European nations are spending—or have spent—the equivalent of hundreds of billions of dollars on efficient public transportation, energy, and water systems. Meanwhile, the American Society of Civil Engineers estimated in 2005 that it would take $1.6 trillion simply to make U.S. infrastructure dependable and safe. The obvious, negative impact of this situation on our global competitiveness, quality of life, and ability to create American jobs is a problem we no longer can ignore.

One way to finance the rebuilding of our country is by creating a national infrastructure bank that is owned by the federal government but not operated by it. The bank would be similar to the World Bank and European Investment Bank. Funded with a capital base of $50 to $60 billion, the infrastructure bank would have the power to insure bonds of state and local governments, provide targeted and precise subsidies, and issue its own thirty- to fifty-year bonds to finance itself with conservative 3:1 gearing. Such a bank could easily leverage $250 billion of new capital in its first several years and as much as $1 trillion over a decade.

Run by an independent board nominated by the president and confirmed by the Senate, the bank would finance projects of regional and national significance, directing funds to their most important uses. It would provide a guidance system for the $73 billion that the federal government spends annually on infrastructure and avoid wasteful "earmark" appropriations. The bank's source of funding would come from funds now dedicated to existing federal programs.

Legislation has been proposed that would create such an infrastructure bank. Congresswoman Rosa DeLauro (D-CT) has introduced a House bill, and Senators John Kerry (D-MA) and Kay Bailey Hutchison (R-TX) have brought forward legislation in the Senate. The Senate bill, with $10 billion of initial funding, is a modest proposal but passing it would give us a strong start.

We should regard infrastructure spending as an investment rather than an expense and should establish a national, capital budget for infrastructure. While this idea is not new, it has been unable to gain political traction. From a federal budgeting standpoint, it would be the wisest thing to do. President Obama and Congress should take action promptly.

Richard Little, Director, Keston Institute for Public Finance and Infrastructure Policy

The massive network of seaports, waterways, railroads, and highways we built in the nineteenth and twentieth centuries were designed to unlock the nation's natural resources, agriculture, and manufacturing strength and bring these products to market. Today, despite a dynamically changing economy, these sectors along with trade and transportation still account for more than a quarter of U.S. GDP or $3.5 *trillion*, but many transport linkages have become bottlenecks due to long-delayed repair and replacement. The entire U.S. economy, as well as consumers, would benefit from a more efficient and resilient supply chain.

Unfortunately, for far too long, Americans have been lulled by their political leadership into a false sense of entitlement. Faced with the prospect of raising taxes or charging fees to cover the cost of maintaining these systems, they have chosen to do neither. As a result, our highways and bridges decline at alarming rates. Most of the other systems vital to our interests suffer the same fate. Fixing this is well within our control, the challenge will be to muster the will to do so.

Without a move to revenue-based models, necessary renewal of critical infrastructure will be long delayed, if provided at all.

The first step in addressing this problem will be to ensure that adequate revenue streams are in place. Whether this revenue comes from the fuel tax, tolls, or other mechanisms is less important than having the funds to work with. Without a move to revenue-based models, necessary renewal of critical infrastructure will be long delayed, if provided at all. We can show that we value these systems by agreeing to pay for their upkeep or own both the responsibility for economic decline and its consequences.

Stephen Goldsmith, New York City Deputy Mayor for Operations

Investment in America's physical infrastructure is directly tied to economic development. Businesses and the workforces they attract consider infrastructure when deciding where to locate. Too often, however, pressed by day-to-day concerns, state and local governments fail to adequately plan and invest in infrastructure. Tight budgets make it easy for officials to rationalize the deferral of investment until a time when surpluses return.

Unfortunately, this pattern has been repeated for decades, and the accumulation of deferred maintenance and deferred investment in future infrastructure has led to an unsatisfactory status quo. To ensure America's future competitiveness in the global marketplace, we must rethink our approach to the construction and financing of infrastructure. And in this policy area, many of the most promising ideas for unlocking public value involve public-private partnerships.

Public-private partnerships can produce access to capital that will accelerate the building of critical infrastructure in sectors ranging from transportation to wastewater treatment.

The key question in a debate about infrastructure should be: "How can we produce the most public value for the money?" Answering this question should lead us to pursue both operational and financing innovations. The private sector has an important role to play in both. Public officials can produce more value for the dollar by better structuring the design, construction, operation, and financing of infrastructure projects that produce more lifecycle benefits and fewer handoffs among various private parties. A private partner can often achieve savings for government by identifying operational efficiencies and assuming risk formerly held by the public sector. Unlike the traditional model for bridge construction in which one firm designs, one firm builds, one company finances, and the public maintains, an arrangement which gives the private firm an ongoing responsibility for maintenance or durability will encourage design optimization and likely increase the length of the asset's lifecycle.

Public-private partnerships can produce access to capital that will accelerate the building of critical infrastructure in sectors ranging from transportation to wastewater treatment. However, maximizing their potential to solve America's infrastructure challenges also requires governments to create a regulatory climate conducive to them. Government agencies should be given maximum flexibility to enter into partnerships with the private sector; and private companies should not have to navigate unreasonable tax laws that limit their ability to partner with government entities to produce better public value.

At a time when every dollar counts, extracting maximum public value out of infrastructure investment is crucial. The private sector can be a strong partner to government. By prioritizing long-term value creation over short-term politics, America can bridge the infrastructure divide and ensure our continued prosperity.

Rebuilding America Special Report[*]

How to Fix U.S. Infrastructure

By Erik Sofge
Popular Mechanics, April 9, 2008

A raw wind is cutting across the Mississippi River, freezing exposed skin at the sprawling construction site for the St. Anthony Falls Bridge in Minneapolis. The project is hurtling toward completion. Three crane drivers work at their controls on the south side of the river, while on the opposite bank four of the machines blend into the towering piers that will support the span. Hundreds of workers labor here in staggered 10-hour shifts, with some crews clocking 60 hours a week.

The urgent pace is the only sign of the collapse of the old I-35W bridge, which occurred on this spot six months ago. The debris was quickly cleared away, and in the aftermath, the Minnesota Department of Transportation (Mn/DOT) pledged to open a new bridge by Dec. 24, 2008. In a business where it can take years just to get a permit, this is an improbably ambitious schedule. "We know that eyes are on us, but that's a good thing," says John Chiglo, Mn/DOT's manager for the project. "Re-establishing public trust is something we felt needed to be done. Something was lost that day, when the bridge collapsed."

And not just in Minnesota. To many Americans, the I-35W disaster wasn't an isolated tragedy, but the latest in a barrage of infrastructure failures—from the northeastern blackout in 2003 and the breached New Orleans levees in 2005 to falling concrete in Boston's Big Dig in 2006. Perhaps the nation had passed a tipping point and was entering a period of steep physical decline.

When the bridge crumbled, Popular Mechanics was midway through a yearlong investigation of the country's infrastructure, focusing on solutions being developed and put into practice right now, at locations such as a massive lock system on the Ohio River, an electric-grid research facility in Washington state and the country's

---

busiest port, located in California. We found promising technologies and innovative projects.

But there are larger lessons to be learned as well. Americans need to face the sobering reality that the country's infrastructure is in trouble. Most of it was built in the 20th century, during the greatest age of construction the world has seen. The continent was wired for electricity and phone service, and colossal projects, including the Hoover Dam, the Golden Gate Bridge and the interstate highway system, were completed—along with thousands of smaller bridges, water tunnels and more. We are living off an inheritance of steel-and-concrete wonders, grander than anything built by Rome, constructed by everyday giants bearing trowels, welding torches and rivet guns.

To fix our infrastructure, from dilapidated levees to congested roadways and ports, the American Society of Civil Engineers (ASCE) has estimated that the country needs to spend $1.6 trillion over five years. Only $1 trillion of that, the organization says, has been allocated or promised. Accepting those numbers, we need an additional $600 billion to reverse the slide of infrastructure, a figure that seems as difficult to produce as it is to comprehend.

Or is it? Spread over five years, ASCE is calling for $120 billion per year. The economic stimulus package signed into law in February is sending $168 billion out to individuals to spend, in a best-case scenario, on new TVs and restaurant meals. That money could have bought a lot of concrete. While more funds are needed, how they're spent is equally important. New information technology, fresh engineering and advanced materials can help us not just restore, but improve our infrastructure in the coming century. Planned and managed properly, next-gen projects can be smarter and more resilient than what came before. Engineers and construction workers know how to get the job done. But first, we must gather the national will.

On March 14, 2006, Ka Loko Reservoir was full. Heavy rains had swamped the island of Kauai, Hawaii, since February, triggering flash floods and road closures. Shortly after 5 AM, the dam breached and 400 million gal. of water exploded into Wailapa Stream, sweeping away trees, cars and two houses. Among the three bodies later identified was that of 22-year-old Christina MacNees, who was seven months pregnant. Four victims, including 2-year-old Rowan Fehring-Dingwall, have yet to be found.

There are more than 83,000 registered dams in the United States, and federal law requires them all to receive regular inspections. When the Ka Loko Dam breached, it had never been examined. A civil probe found that the dam's owner had been uncooperative—inspectors made two attempts but never gained access. They may have been lulled by a bureaucratic blind spot: Years before, Hawaii had classified the dam as "low-hazard," implying that even if it failed, lives wouldn't be at risk. "It's called hazard-classification creep," says Dan Johnson, a civil engineer with GEI, a geotechnical consulting firm that specializes in water resources. "When a lot of dams were built, they were considered low hazard. If one failed, it would maybe wash out an antelope. But today, that's a city."

In addition, the database of dams ignores some low-probability/high-consequence events, according to Victor Saouma, a professor of civil engineering at the University of Colorado in Boulder. "If a dam were to collapse following an earthquake in California, we're talking about thousands of people dying," Saouma says. Researchers haven't determined which dams are in greatest danger, let alone how to fix the problem.

This is an age in which banks track millions of accounts on a second-by-second basis. Ordinary people can call up satellite images of every corner of the globe in seconds, day or night. So it's remarkable that the victims of Ka Loko Dam died—and more people could die—because engineers lack up-to-date data. (And it's not just dams. Three years after Hurricane Katrina, there's no central registry of the nation's levees. It's unknown how many thousands of miles of earthen walls may be on the verge of collapse.)

There is a better way. Hawaii is working with the Pacific Disaster Center to develop computer models for predicting the consequences if a dam were to fail. Such models, which incorporate weather and mapping data, should be employed nationwide. New technology can help—the Army Corps of Engineers has begun using remotely operated vehicles to inspect underwater structures. Other tools have been developed but aren't yet in use as broadly as they could be. These include GPS sensors to detect subtle shifting of a dam's structure, and geographic information systems to make it easier for officials—and perhaps concerned citizens—to access the data.

Fixing the country's dams will also take a sense of urgency. Kentucky's Wolf Creek Dam, upstream from Nashville, Tenn., is considered one of the most dangerous in the country. In 2005, the Army Corps of Engineers noticed an alarming amount of seepage under and around the dam, despite major repairs in the 1960s and '70s. A permeable limestone foundation supports the structure; it dates from 1952, and engineers say the design wouldn't be approved today. Crews have begun adding concrete grout to the foundation (about 500,000 gal. by early 2008) and installing an additional concrete wall. However, it remains an open question whether the dam can hold out until the project is completed in 2012.

For two weeks in July 2006, California baked in a heat storm of unrelenting, 100-degree-plus temperatures. Air conditioners ran day and night, overtaxing the electrical grid. On July 24, when power demand hit more than 50,000 megawatts—the highest level in state history—transformers started failing. Utility Pacific Gas & Electric was quickly overwhelmed, and more than a million people lost power, some for days. When the heat finally broke, it was blamed for 141 deaths.

"Our grids today are more stressed than they have been in the past three decades," says Kevin Kolevar, assistant secretary for electricity delivery and energy reliability at the Department of Energy. "If we don't expand our capacity to keep up with an increase in demand of 40 percent over the next 25 years, we're going to see healthy grids become increasingly less reliable." Today, with the grid operating flat-out, any disruption—like the downed transmission line that sparked the 2003 blackout in the Northeast—can cripple the network.

Building more power plants will give grid operators a larger safety margin, and new technologies can make it easier to exploit renewable energy sources such as wind farms and solar arrays. "Wind power is erratic. It goes when the wind blows, which is often when demand is at its lowest," says Jim Detmers, vice president of operations for California ISO, which directs the state's grid. "We need new tools to manage that flow." One answer is to incorporate systems to store the excess power, instead of feeding it directly into the grid. Advanced batteries, such as new Japanese-built sodium-sulfur models, can store 5 megawatts. The grid in Fairbanks, Alaska, is backed up by the world's biggest battery, a 40-megawatt NiCd (nickel-cadmium) system that came online in 2003—and was called into service 82 times in 2006 alone. Flywheels are being developed to provide instantly accessible stored energy, while ultracapacitors can provide "wind-power smoothing."

Cristina Archer, a professor of civil and environmental engineering at Stanford University, has proposed a different approach. By analyzing data from existing wind farms in the Midwest, Archer and a partner found that turbines spinning at different times can help fill gaps in the supply. The study determined that if 19 of those farms were networked together, grid operators could safely rely on more than a third of the total power generated—it would become as dependable as electricity from a traditional power plant.

End users may help, too. Last year, researchers at Pacific Northwest National Laboratory (PNNL) conducted a pilot program in Washington state in which households were paid to let grid operators temporarily reduce the power to their homes. The idea was to simulate a peak-demand situation, when even a 1-second drop in power usage can prevent a blackout. A similar test involved clothes dryers designed by Whirlpool to cycle off in response to conditions in the grid. The idea is being tried with other appliances, as well—imagine a refrigerator compressor that delayed kicking on for a few minutes when demand was spiking.

"It creates a shock absorber for the power grid," says Rob Pratt, who led both exercises as the head of PNNL's GridWise research program. "You cushion the blow with demand and give grid operators time to respond." The results: a 15 percent reduction in total demand through real-time incentives and a 20 percent drop through smart appliances. Adopted nationally, such innovations could be revolutionary. They might eliminate the need for at least some new coal-fired power plants—and they would make the grid far more resilient, able to handle adversity without going into a tailspin.

At 6:05 PM on Aug. 1, 2007, several lanes on Minneapolis's I-35W bridge were shut down for repaving, and rush-hour traffic was at a crawl. Peter Hausmann, a 47-year-old computer security specialist, was on the phone with his wife when the main span collapsed, pouring his car into the water 108 ft. below. He was among 13 people killed; 145 more were injured.

"The I-35W bridge was labeled 'structurally deficient' for 16 years," says Jesus de la Garza, a professor of civil and environmental engineering at Virginia Tech. "That meant it was due for reconstruction at some time in the future." Had the work taken place, the catastrophe wouldn't have occurred. (According to a preliminary finding of the National Transportation Safety Board [NTSB] released in January 2008, steel gusset plates had fractured, owing mainly to a flaw in the original design.) There are 153,521 structurally deficient or functionally obsolete bridges in the United States, and, like the I-35W bridge, many remain on the nation's to-do list for decades. "The crisis is deferred maintenance," de la Garza says. "We have too much to build and not enough funding."

However, money alone isn't enough. In July 2006, Milena Del Valle, a 38-year-old mother of three, was killed by 26 tons of falling concrete in Boston's just-completed Big Dig roads project. Funding wasn't the issue—the project had run up massive cost overruns, bloating from a projected $2.6 billion to an eventual $14.8 billion. The problem? According to the NTSB, a contractor used an inappropriate epoxy for the ceiling tiles. It was one more blunder on a project where a confusing web of contractors and subcontractors seemed to undermine decision making and accountability. "The culture was not to have a large number of people, inside or outside the organization, ask questions," says David Luberoff, executive director of Harvard's Rappaport Institute and co-author of Mega-Projects: The Changing Politics of Urban Public Investment. "Management and culture really matter."

Elsewhere, contractors and government officials are finding more effective ways to manage projects. In April 2007, after a fuel tanker truck burst into flames below a congested overpass in Oakland, Calif., triggering a massive collapse, authorities expected repairs to take 50 days. The contractor, C.C. Myers, Inc., claimed it would finish in half that time, to secure the maximum early completion bonus of $5 million. What followed was a master class in efficiency, as the firm took financial risks, such as racking up overtime costs, while showing a rare level of collaboration with subcontractors and transportation officials. The result: The overpass reopened just 17 days after work started, or 33 days ahead of the deadline.

Virginia's Department of Transportation has turned to privatization to improve its highway maintenance. Instead of providing a detailed plan on how to repair a stretch of road, officials assign performance targets, and then ask firms to assume ownership of the route for as long as 10 years. "You don't tell the contractor how to maintain these roads and bridges," says de la Garza, who has helped the state develop the initiative. "The contractor has to be innovative in how they go about maintaining performance."

Innovation and accountability were built into the St. Anthony Falls Bridge project in Minneapolis at the outset. Normally, officials determine the design of a new project before taking construction bids. To replace the I-35W bridge, Mn/DOT opted instead for a "design-bid" approach. Teams provided their own plans, along with guaranteed time frames and detailed engineering credentials. The winning team—Flatiron-Manson and FIGG—didn't submit the lowest price or the shortest schedule, but it did demonstrate the most know-how.

In winter, the warmest place to watch the bridge come together is from inside a series of five-story-tall hangarlike structures where crews are constructing massive sections of the span. Heaters keep air temperatures at 40 degrees to help the concrete cure properly. When one section is finished, workers slide the house down the line, leaving a trail of monoliths in their wake. Everything seems to happen simultaneously, and that's been true since work began. When test shafts were drilled in November, engineers realized they'd need to adjust the placement of the footings by 5 ft. On another project, getting approval might have taken weeks; here, they made the appropriate change and kept moving.

Outside the warming houses, a high-pitched buzz pierces the whine of generators and heavy equipment. Against the flat winter sky, a black dot grows bigger—an unmanned aerial vehicle is coming in for a landing in a nearby field. This toy-size radio-control plane has been on duty since construction started in the fall, capturing aerial images that the contractor uses to help stay on schedule.

The work site is full of such innovations. There's an old engineering adage: You can have it good, fast or cheap—pick two. At a projected cost of more than $230 million, the St. Anthony Falls Bridge will not be cheap, but more than money is at stake. Plans call for high-performance concrete, a structural monitoring system and a life span of 100 years. If it comes in on time, the bridge will prove what engineers and workers can do when they have the freedom to innovate and the resources to get the job done. On the riverbank where everything fell apart, a model for building the future may be coming together. America is watching.

GOAL: FIX THE LEVEES

ACTION PLAN: Federal officials oversee more than 14,000 miles of levees that protect homes an property. No one knows how many thousands of miles are in private hands. Many levees are earthen structures built more than 50 years ago, which are now in danger of collapse. In a positive move, officials are now using lidar, a laser-based equivalent of sonar, to examine underwater structures. Inspections should be stepped up, and the most dangerous levees repaired.

1. SEEPAGE BERM

A heavy layer of sod or dirt combats under-seepage by reducing boils.

2. SLURRY CUTOFF WALL

Levees can be retrofitted with a 2- to 3-ft.-wide slap of concrete up to 120 ft. deep to stop under-seepage. Alternately, steel sheet piling may be used.

3. RAISED LEVEE

Levees settle over time; adding earth restores freeboard, the structure's clearance above the waterline. It also reduces seepage through the levee.

4. SHRUBBERY

Low vegetation on the water side of a levee can guard against erosion (and provide wildlife habitat).

5. BANK PROTECTION

Layers of soil and rock serve as a buffer against fast-moving water that can undercut the levee walls

A recent levee failure in Nevada may have been caused by burrowing rodents. Other common problems: roots from poorly placed trees that can weaken the structure, erosion, and under-seepage through weak foundations, which can produce a spring, or "boil." In 2004, a levee breach near Stockton, Calif, threatened state freshwater supplies.

SACRAMENTO-SAN JOAQUIN RIVER DELTA, CALIFORNIA

PROGRESS REPORT: More than a thousand miles of eroding levees in the Sacramento-San Joaquin River Delta protect the freshwater supply for two-thirds of California residents. Left alone, the levees will fail. One solution proposed by researchers at the Public Policy Institute of California: Build a separate structure to carry fresh water through the region, and let the natural flood cycle return to portions of the delta.

GOAL: FIX THE GRID

ACTION PLAN: The United States uses 4 trillion kilowatt-hours of electricity each year, and the figure is expected to climb, outstripping our generating capacity. More power plants are only part of the answer. Using networking technology to monitor—and react to—what's happening in the grid at each moment can improve efficiency and prevent outages. Decentralizing the production of electricity can also make the grid more resilient and save some of the 400 billion kwh now lost while current flows through long-distance transmission lines to the nation's households.

1. SMALL-SCALE POWER

Microgrids will be small areas—like the residential and industrial neighborhoods shown here—where energy needs are roughly matched by local generation. A control station will juggle demand, buying and selling power to the main grid. During a regional blackout, a microgrid can run in "islanded" mode.

2. ADVANCED MONITORING

Today, utilities don't learn of local power outages until they receive angry phone calls. Roger Anderson, a researcher at Columbia University's Lamont-Doherty Earth Observatory, has worked with Con Ed to develop a system that uses real-time data and machine-learning techniques to help the utility prevent outages. "We want to be predictive instead of reactive," Anderson says.

3. COGENERATION

About 60 percent of the energy used to generate electricity in power plants is wasted as heat. A more efficient approach: Install small, natural-gas-powered cogeneration units in individual buildings to make electricity, and capture the heat for climate control.

4. ELECTRICITY STORAGE

Plug-in hybrid cars could give the grid a power reserve. Drivers would charge the cars at night, then sell some power back during the day.

5. DEMAND PRICING

Spikes in electricity usage can damage transformers, triggering blackouts. Studies run by the DOE's Pacific Northwest National Laboratory showed that with the right pricing and technology, homeowners could buffer the grid from trouble by letting it temporarily reduce their incoming power load.

6. TRADITIONAL POWER

The United States will need more big power plants—and added transmission lines to connect them with users. High-temperature superconducting transmission lines can carry three to five times the current of copper lines, while superconducting transformers can cut transmission losses in half.

GOAL: FIX THE BRIDGES

ACTION PLAN: One-quarter of the 599,893 bridges in the United States have structural problems or outdated designs. The country can do more than rebuild these bridges—we can make them better, using high-performance concrete, steel and composites; automated monitoring systems to watch for deterioration; and smarter designs. Similar technologies can also be employed on highways, tunnels and other structures.

1. SUPER STEEL

High-performance steel has been used since 1998. A new version can retain its integrity to 100,000 psi of pressure, twice the level of conventional steel. Lehigh University researchers are using the steel to re-engineer I-beams with corrugated webs, making them thinner but stronger than previous beams.

2. FIBER-REINFORCED POLYMER

Today's rebar-and-concrete bridge decks are vulnerable to weathering. An alternative, fiber-reinforced polymer (FRP) grids, has been built into dozens of demonstration projects in the United States. The oldest is 11 years and shows no sign of corrosion. More than 300 bridges incorporate other FRP components.

3. MORE-DURABLE ROAD SURFACES

Conventional road surfaces can last 40 years with regular maintenance. A new concrete developed at the Missouri University of Science and Technology may last 100 years and withstand up to 30,000 psi of pressure—nearly eight times as much as ordinary material.

4. SELF-HEALING MATERIALS

Structural problems often begin as microscopic fissures. When a crack begins in a fiber-reinforced bendable concrete developed at the University of Michigan, material newly exposed to the elements cures into concrete that's as hard as the original.

5. SENSORS

Periodic inspections by officials can be complemented by permanent sensors. The Smart Brick, a wireless system from the Missouri University of Science and Technology, detects flood conditions and measures temperature, vibration, humidity and more. A sensitive "skin" created at the University of Michigan alerts engineers to deformations and widening cracks by tracking changes in electrical conductivity.

6. FRP PILINGS

Typically, to make pilings, workers drive steel forms into the ground, then fill them with concrete and rebar. To create more durable, corrosion-resistant pilings, workers can pour concrete into FRP forms at a factory. The units are then brought to the site and driven into the ground. (Diagram by Axel De Roy)

ST. ANTHONY FALLS BRIDGE, MINNEAPOLIS

PROGRESS REPORT: Minnesota officials pledged to quickly replace the collapsed I-35W bridge. The new span, due to open on Dec. 24, 2008, is being built using high-performance concrete and 15.8 million pounds of rebar, much of it treated to thwart corrosion. Sensors will watch for structural problems. The bridge will be 1223 ft. long and carry 10 lanes of traffic. It is designed to last 100 years.

GOAL: FIX THE PORTS

ACTION PLAN: About 28.9 million shipping containers passed through crowded U.S. ports last year, and gridlock is mounting. Containers entering the country languished on docks an average of seven days. Adopting the "agile port system" now being developed with help from federal agencies would boost efficiency. When the concept was tested at Washington's Port of Tacoma, it cut cargo delays in half. ports

1. SMARTER CARGO HANDLING

Cargo unloaded from ships must be sorted for transport by train or truck. RFID technology and optical scanners can help cut this step: Cargo would be grouped by destination at the point of origin. Using multiple cranes to simultaneously load and unload different ship holds can boost efficiency. New electric-powered machinery on the docks can reduce air pollution.

2. DEEPER CHANNELS

Dredging 45-ft.-deep channels allows super-post-panamax ships—which carry more than 8000 containers—to access more ports, easing congestion in the busiest locations.

3. CLOSER RAIL LINES

Containers typically are driven from ships to rail yards less than 5 miles away. Extending tracks onto wharves can save time and fuel. The cargo might travel by rail to inland ports before moving on to market.

4. ELECTRIC HOOKUPS

A container ship running its auxiliary engines to generate electricity can burn 1600 gal. of diesel fuel each day while in port. Electrical hookups that let ships shut down their engines aren't part of the agile port concept, but they can sharply reduce air pollution. A study of the Port of Long Beach indicates that a container ship that visits the city eight to 10 times annually could cut its harmful emissions by 76 tons each year.

PORTS OF LOS ANGELES AND LONG BEACH

PROGRESS REPORT: In 2007, California's twin ports of Long Beach and Los Angeles received more than 8000 vessel calls and handled 15.7 million containers, making the complex the country's largest shipping gateway—and the fifth busiest in the world. Traffic congestion and air pollution are major concerns of local residents. To reduce truck traffic, the ports helped develop a 20-mile dedicated rail corridor that carried 4.7 million containers last year. The ports are laying rail line onto docks to further reduce the need for trucks, and they plan to operate 25 electric-ready berths by 2011.

GOAL: FIX THE LOCKS

ACTION PLAN: Navigable waterways are America's stealth transportation system. "We're aware of trains when we're sitting at the crossing and of trucks when they're flying by on the highway," says Mark Hammond, an economist with the Army Corps of Engineers. "But we tend not to take notice of the river, which carries an awful lot of stuff." The country's more than 12,000 miles of inland waterways transport 625 million tons of freight each year, including coal and grain. About half of all U.S. lock-and-dam systems need to be replaced or modernized.

The new Olmsted Lock and Dam has two 1200-ft.-long, 110-ft.-wide chambers. Once barges enter, operators will close the gates and adjust a set of valves to raise or lower the water level as needed. No pumps are required.

OLMSTED LOCK AND DAM, ILLINOIS

PROGRESS REPORT: The Ohio River's Olmsted Lock and Dam, located on the border of Illinois and Kentucky, will replace rotting, 1920s-era structures. Along with other cargo, 12 million tons of coal travel through the current locks by barge each year, fueling power plants that supply electricity to millions of customers. (It would take 460,000 semi trucks to replace these barges.)

Olmsted's lock chambers are being reinforced with layers of wrist-thick steel rebar to withstand possible earthquakes in the New Madrid Fault. Approach walls will float instead of being fixed to the river bottom, saving building and maintenance costs. Modern structures to control water levels will substitute for the current locks' timber wicket gates, which are operated by a dam tender in a boat.

A New Bank to Save Our Infrastructure[*]

By Felix G. Rohatyn and Everett Ehrlich
The New York Review of Books, October 9, 2008

I.

These are rare times of ferment in one of the most neglected fields of public policy—the nation's infrastructure, or what used to be known as public works, including roads, mass transit, bridges, ports and airports, flood control systems, and much else. We have been confronted with spectacular and tragic evidence of the inadequacy of these facilities in the failure of the levees in New Orleans and in the collapse of the I-35 bridge in Minneapolis. More generally, a recent report by the American Society of Civil Engineers concludes that America's infrastructure overall is close to "failing" and deserves a grade of "D." It estimates that an investment of $1.6 trillion will be needed to bring it up to working order.

According to the report, nearly 30 percent of the nation's 590,750 bridges are "structurally deficient or functionally obsolete" and it will take "$9.4 billion a year for 20 years to eliminate all bridge deficiencies." "The number of unsafe dams has risen by 33 percent to more than 3,500." Public transit facilities—including buses, subways, and commuter trains—are dangerously under-funded, even as demand for them has "increased faster than any other mode of transportation." Current funding for safe drinking water amounts to "less than 10 percent of the total national requirement," while "aging wastewater management systems discharge billions of gallons of untreated sewage into US surface waters each year." Yet government investment in these vital facilities is generally held to be below the level needed simply to maintain them in their current poor state.

The gap between our economy's need for functioning infrastructure and what is being invested in it has aroused much concern. Tired of waiting for Washington to recover the vision and energy it once devoted to the problem, Governor Arnold Schwarzenegger convinced California voters in 2006 to approve $20 billion in bonds to finance the repair and construction of roads and bridges in the state as well as public transit systems and other facilities. Together with Governor Ed Rendell of Pennsylvania and New York Mayor Michael Bloomberg, Governor Schwarzenegger has also formed a bipartisan group called Building America's Future, which aims to find better ways to address the crisis. A second group, the Transportation Transformation Group, led by, among others, former House majority leader Dick Gephardt and General Barry McCaffrey, former US Southern Forces commander and drug czar, has a similar mission and the backing of Goldman Sachs.

Along with the Australian company Macquarie, Goldman Sachs is also among a new group of investors who are taking part in private refinancings of toll roads such as the Chicago Skyway, the Indiana Toll Road, and now perhaps the Pennsylvania Turnpike and the New Jersey Turnpike. Under those arrangements, the state or city sells the road and the right to set and collect tolls on it to a private company—in essence, a new form of government borrowing.

The last element of this mounting interest in the problem of infrastructure is public frustration at the costs to consumers of poorly maintained roads, bridges, transit systems, and airports. The average American motorist incurred $710 in lost time and fuel costs in 2005, well before the price of oil went over $100 per barrel. Air travelers fare no better—there were 1.8 million hours of flight delays in the US in 2007, many of which were caused by demands for runways that exceeded supply. Shippers report increasing frustration with the nation's ports. According to the American Society of Civil Engineers, it will take over a quarter of a trillion dollars to bring the nation's public school buildings up to "good" condition. And the demand for all of these services will increase further with population growth and economic activity.

But while private investors and states and cities are devoting more attention to this, the federal government has failed to provide the leadership it alone can supply. Federal spending on infrastructure, corrected for inflation, is actually lower than it was in 2001, despite the growing economy, the well-known disrepair and obsolescence of our assets, and the rising costs of their inadequacy. And this level of spending, as a share of GDP, is much lower than it was two or three decades ago.

Throughout US history, competent public investment decisions have been an essential complement to private investment, from the Louisiana Purchase and the Land Grant Colleges to the Interstate Highway System and the Internet. And the functions of infrastructure are still as essential as they have ever been, if not more so. Indeed, *The Economist* reports that China will spend $200 billion on its railways between 2006 and 2010—the largest investment in railroad capacity made by any country since the nineteenth century—while the US rail system continues to become more and more degraded at a time of great potential renewal. The Chinese also plan over the next twelve years to construct 300,000 kilometers of roads in rural China, as well as ninety-seven new airports. The Chinese understand that economic power depends on these investments.

In an effort to confront this problem, Congressman John Mica, the ranking Republican member of the House Transportation Committee, recently called for a trillion-and-a-half-dollar infrastructure spending program, under both public and private sponsorship. But where would the money come from? The Iraq war drains our national resources, and the 2001 cuts in personal income, capital gains, and inheritance taxes have slashed federal revenues. Meanwhile, several presidential candidates, including the Republican nominee, Senator John McCain, were unable to resist the temptation to endorse a motor fuels tax "holiday," which would produce negligible saving for motorists but cut even further needed federal revenues. Thus, when it comes time for investments in our future, the federal cupboard is bare.

This public penury is lamentable, but it conceals a second and perhaps even more fundamental problem with federal policy: not only do we fund infrastructure inadequately, but the policies we have in place are incapable of funding the needed projects or creating the incentives to manage correctly what's already been built. This is the unseen and ultimately more critical part of the infrastructure crisis— the extent to which our spending programs are misdirecting our investments away from the best opportunities.

2.

Responsibility for the nation's infrastructure is currently spread across federal, state, and local governments. For example, the federal government is responsible for maintaining wastewater systems, while states and municipalities handle drinking water. The federal government helps states, cities, and towns build and operate mass transit systems; and it builds bridges that are part of the Interstate system, while local governments build local roads and the picturesque covered bridges that appear on tourist postcards. Most of the federal government's $73 billion budget for infrastructure in 2007 was spent on a handful of "modal" programs dedicated to promoting the construction and major rehabilitation of specific types of infrastructure, or "modes"—the Federal-Aid Highway Program, the Airport Improvement Program, the Transit Formula and Bus Grant Program, and the Army Corps of Engineers' water resource programs, among others.

While details of these programs vary, their basic workings are similar. States and cities propose projects to each of these "programs." The federal government then decides which projects to pursue, either funding all or most of the cost of those projects, or sending blocks of money to state capitals (as does the Highway Trust Fund), where state governments dole it out. Except for the Corps of Engineers, however, infrastructure program officials administer grants rather than carry out construction and other work.

Some projects (often navigation or water resource development) benefit from selective congressional patronage—either so-called "earmarks" (special bequests for the pet projects of specific representatives or senators, such as the infamous Alaskan "bridge to nowhere") or deals cut between Congress and the agencies. These deals are typically used for such water projects as the St. John's Bayou–New Madrid Floodway Project in southeast Missouri, described by the corps's own officials as an "economic dud with huge environmental consequences" and "a bad project. Period."[1]

In the first part of the twentieth century, the nation was still developing highway and airport systems, and these methods of financing worked. States were eager to get federal funds to integrate their roads and airports into the new national networks. But that job was substantially done by the 1980s, and we now find ourselves, as General Heinz Guderian remarked, fighting the next war with the tools of the last one. Sending federal money to state capitals to fund 90 percent of whatever road construction state legislatures choose does little to further projects of national scope or genuine economic value. Moreover, the availability of funds to build new roads often blunts the incentive to repair and maintain existing roads until their deficiencies become pressing enough to warrant reconstruction. Thus, little has been done to maintain the Interstate Highway System, despite the fact that major sections are falling into disrepair, and repairing them is estimated to provide the taxpayers with the highest economic returns among highway projects today—much higher than the returns from building new roads.

Federal grants for water projects create other disincentives. If the Corps of Engineers doesn't get around to funding a city's project, then that city has every reason to wait for the next budget cycle instead of looking for other solutions. The result is lethargy and delay. The federal government will typically pay for levees, but not to preserve wetlands that provide natural flood protection by absorbing torrential rain. Moreover, by shielding local users from the true cost of living on flood plains, federal programs encourage development in areas that cannot sustain it.

Hurricane Katrina demonstrated the potentially devastating consequences of these failures. The state of Louisiana and its municipalities built flood control systems around levees while ignoring the deterioration of fragile wetlands in the Mississippi Delta. Louisiana's congressional delegation steered federal funds toward navigation projects instead of flood control.

Particularly unfortunate is the failure of government to consider alternatives to new infrastructure construction. As the residents of any major American city understand, there are few places left to build new roads to relieve urban congestion or to expand or build new airports to reduce delay. Sooner or later, the officials in charge must consider managing road and airport use through pricing. They may auction off landing slots during peak rush hour periods[2] or reserved lanes for drivers paying congestion tolls, much as we have "high-occupancy" lanes today. Or, as Mayor Bloomberg bravely proposed in New York, they may impose fees for bringing an automobile during business hours into the most congested part of Manhattan, a program that has been enormously successful in London.

In a world of $4-a-gallon gas and $40,000-a-year college, raising tolls will be unpopular with many families, whether poor or well-to-do. But we must either accept congestion and delay—together with ever more deteriorating and dangerous infrastructure—or use tolls to limit public use while providing a new source of revenue for transportation improvements. London now realizes about $400 million annually from the fees it charges to drive into the city; and it has used these revenues to expand dramatically its bus fleet. New York's revenue would likely be higher, and would support any number of mass transit programs that would benefit lower-income users and commuters generally.

Another consequence of having different government programs dedicated to different types of infrastructure—whether highways, water projects, or wastewater treatment—is the creation of bureaucratic fiefdoms that are inevitably held captive to the "iron triangle" of congresspeople, lobbyists, and the bureaucrats themselves, as has happened in the case of the Highway Trust Fund and the Army Corps of Engineers. As a result, these programs never compete with one another. No responsible body has the mission of impartially deciding whether we'd be better off with more mass transit and better train service and fewer major roads, because these are never compared when a specific proposal is under review. Moreover, the different agencies that analyze projects—if they do so—generally use different (and self-interested) criteria for determining such critical variables as the value of time, the value of new jobs created, the discount rate, the cost of capital, and so on.[3] As a result, the public is left without the apples-to-apples comparisons that any rational investor would use to allocate a portfolio of billions of dollars of investment.

So the "modal" infrastructure programs, rather than competing efficiently for resources, all lurch forward without coordination or attention to the merits of the specific projects they choose to fund. And that is in cases when the programs are not directly muscled through by politicians. The term "earmark" became popular during the writing of the 2005 transportation bill, which contained over six thousand of them (with a total cost of $24 billion), compared to five hundred of them in 1991 and ten in 1982.

3.

In view of the waste and inadequacies of existing federal and state policies, how can we begin to address the growing infrastructure crisis? In September 2004, former Senator Warren Rudman and one of the authors of this essay, Felix Rohatyn, agreed to chair a Commission on Public Infrastructure at the Center for Strategic and International Studies (CSIS) in Washington, D.C., to outline a new and different approach to selecting, financing, and managing infrastructure. Last year, the commission produced a consensus report; and a bill to enact its approach, the National Infrastructure Bank Act of 2007, has been submitted by Senators Chris Dodd (D., Connecticut) and Chuck Hagel (R., Nebraska), both of whom served as members of the CSIS commission. A companion bill has been offered in the House of Representatives by Banking Committee Chairman Barney Frank (D., Massachusetts) and Representative Keith Ellison (D., Minnesota); while a similar approach has been proposed in a bill introduced by Representative Rosa DeLauro (D., Connecticut). Barack Obama has spoken of the need for "a National Infrastructure Reinvestment Bank that will invest $60 billion over ten years. . . . The repairs will be determined not by politics, but by what will maximize our safety and homeland security; what will keep our environment clean and economy strong."

The central idea of the CSIS commission proposal is to establish a National Infrastructure Bank, an institution that would be similar to the World Bank, a private investment bank, or any other entity that evaluates project proposals and assembles a portfolio of investments to pay for them. Traditionally, public financial institutions such as the one we propose are created to correct problems in capital markets, whether they be the failure of markets to fund projects that support development in the world's poorest nations or their undue pessimism regarding the long-term solvency of a particular city or state government. This is not the case here. State and local governments generally can borrow for infrastructure purposes in line with their ability to service debt and the strength of their credit ratings. The issue here is not the efficiency of capital markets but rather the efficiency with which federal programs work and spend funds. The purpose of the National Infrastructure Bank would be to use federal resources more effectively and to raise additional funding. We propose this bank because we believe that markets for capital do work and can be harnessed to solve the critical shortfall in funding infrastructure.

The bank would replace the various "modal" programs for highways, airports, mass transit, water projects, and other infrastructure, streamlining them and folding them together into a new entity with a new culture and purpose. Any project seeking federal participation over a set dollar threshold would have to be submitted to this bank. (Smaller projects would be left to states, cities, and towns, perhaps with an accompanying federal grant to be used at the discretion of the state or local government.) Rather than receiving grants through pre-set federal formulas or privileged congressional payments, states, cities, or other levels of government would come to the bank with proposals they wished to pursue. These proposals—for, say, a new or improved highway, a subway, expanded airport, or harbor improvements—would outline the investment that state and local governments would be willing to make, what the users of the project would be expected to pay, and what support was wanted from the federal government.

The bank would have no preconceived, overarching plan for the nation's infrastructure. Proposals for infrastructure investment would still predominantly come from state and local governments. Our plan would preserve almost entirely the existing balance of power between federal, state, and local government, but would change dramatically the way priorities are set and projects funded. That is because it would proceed project-by-project, and dollar-by-dollar, to find the best use of federal resources.

The bank would have a board of directors that included key Cabinet officers and members appointed by both the executive branch and congressional leadership; its chief executive would be appointed by the president and confirmed by the Senate. The Federal Reserve, the Public Company Accounting Oversight Board, and the Pension Benefit Guarantee Corporation are all good examples of comparable agencies with expert and important missions that have consistently functioned well. The bank would require states, cities, or other sponsoring entities to seek federal assistance only after they have thought through alternatives such as tolls and other user charges, such as the adjustment of prices to peak loads on the roads and airports or the availability of other solutions that do not require new, burdensome structures. These would include wetlands for flood control or changing speed limits and the use of "smart" traffic systems that allow more cars to use the same limited road space more efficiently. The bank would be in a good position to ask whether applicants were aware of alternatives and had considered the most efficient technology.

Imagine, for example, that the bank received a proposal from a state for a new highway segment and found, using its consistent analytic approach, that the plan had legitimate national benefits. It could then provide support in several ways. It could simply write a check to the state building the road and provide a direct subsidy for some portion of the total cost. Alternatively, it could purchase credit guarantees for the state bonds that financed the roads; or it could provide interest rate subsidies to reduce the rate paid on those bonds. It could lend the money directly to the state and be repaid from tolls; or it could provide sinking funds (funds sometimes set aside to guarantee the repayment of the bond), or underwrite the state's bond offering (guaranteeing that all of the bonds will be purchased at a predetermined price), or take other steps. States and municipalities, of course, could continue to borrow from public markets as they do now; what would change is the federal government's financing role.

The bank's ability to sell securities based on its infrastructure projects such as roads and bridges would also resolve a major quandary of infrastructure policy—how to manage the influx of private money into particular projects. State and local governments too often sell highways and other transportation networks to private investors because they have been unable to raise tolls to sufficient levels, and as a result they risk selling these on the cheap or other bad terms. San Diego has approved a plan to let a private company build a private toll road with the promise that no other road would compete with it for the indefinite future. Chicago's lease of its Skyway road system to a partnership of the Spanish firm Cintra and Macquarie will last for ninety-nine years, far longer than the road itself will! And if local governments use the receipts of such one-off sales for "rainy day" funds or other operating expenditures, they are making their long-term fiscal situations worse, not better.

Although private investors have successfully built new roads in places such as Poland and Spain, they have not done so extensively in the US. But a National Infrastructure Bank could redirect private efforts away from refinancing old facilities—as in the case of Chicago's Skyway—to building new ones. According to our plan, most of the funds the federal government now spends on existing programs (along with many of those program's experts and facilities) would be transferred to the bank, which could not only finance the projects but also resell the loans it makes to investors in capital markets, much as other assets are rebundled for investors. The receipts from these sales would allow a new round of lending, giving the bank an impact far in excess of its initial capitalization. Moreover, selling the loans it makes to private investors would require the bank to convince those investors that its projects are tenable and capable of producing tangible benefits—in short, the bank's project selections would face a market test every day, as a deep and liquid market for its securities was formed. Or, alternatively, the bank could issue its own fifty-year bonds, backed by its loan portfolio, to obtain its own capital.

Even with a conservative ratio of borrowed funds to capital of three to one (meaning each dollar of federal activity attracts three added dollars of private borrowing), this could produce almost a quarter-trillion of investment on a $60 billion annual bond issue. But regardless of the particular financial mechanisms chosen, a freestanding bank would permit raising additional money by borrowing on the basis of the bank's balance sheet and financial capacity. As a result, the bank could produce substantially more investment and hundreds of thousands of new jobs in the first several years of its operation.

The bank's securities, whatever they may be, should not benefit from a promise of the government's full faith and credit (as has been enjoyed and abused by Fannie Mae and Freddie Mac). Only close scrutiny by investors can provide the kinds of discipline needed to ensure the bank's long-term success. If the bank wishes to support a proposed project—whether by writing a check, insuring a local bond, providing other credit guarantees, or lending its own money—its securities should each be carefully exposed and specifically targeted, allowing participating investors to evaluate the assets they buy. But in our view the dramatic need for additional infrastructure investment clearly justifies tax-free returns for those securities.

4.

A final question concerns paying for this new infrastructure policy. As we noted above, the first source of financing should come from the funds now dedicated to existing infrastructure programs—about $60 billion annually could be taken from these programs with a balance left over. And there is nothing wrong with continuing to charge users a motor fuels tax, an air ticket surcharge, port fees, and other fees that now are imposed for using infrastructure. But two further points should be made.

First, we can increase our investments in infrastructure and still have fiscal discipline. There is no shortage of options for raising revenue for investment purposes while still making the tax system more efficient and fair—two examples are a consumption or value-added-tax (perhaps partially offset by lower income taxes to maintain progressivity) or a carbon tax or energy tax. And since it would target its subsidies more effectively, the bank would get more investment out of existing budgetary resources while adhering to the "pay as you go" (PAYGO) budget rules used by Congress, which call for each new dollar of spending to be offset by a dollar of reduced spending or increased revenue elsewhere. At the same time, the bank's financial statements would take us one step closer to having the information that a capital budget would provide—most critically, whether we are investing in infrastructure faster than it is depreciating or becoming obsolescent.

The second point is the matter of fiscal stimulus. Bloomberg, Rendell, and Schwarzenegger have recently urged that increased spending on infrastructure be the center of a new stimulus package, as have House Speaker Nancy Pelosi and former Treasury Secretary Lawrence Summers.[4] This is an attractive prospect—an additional $40 billion in infrastructure investment could create as many as a million new jobs. We share this objective,[5] but believe the best way to accomplish it is through an immediate revenue-sharing grant to states and cities for these purposes. In the interim, a bank along the lines described here and in the Dodd-Hagel bill could be set up and put into operation within a year.

Ultimately, we face a future of mass transit strained beyond capacity, planes sitting on tarmacs, slow traffic and wasteful sprawl, ports that lack the capacity to operate efficiently, and increasing numbers of bridges and dams that are obsolescent and dangerous to the public's health and safety—in short, the dire prognosis of the American Society of Civil Engineers is coming true. Regardless of the government's fiscal position, vital investments in transportation, water supply, education, and clean energy are necessary to maintain our future standard of living. Our political system pours money into war and tax breaks while relying on deficit finance. Those in charge then announce that there are no resources left to secure our economic future. The new bank we propose offers one alternative to such a dangerous set of policies.

FOOTNOTES

1. Michael Grunwald, "Par for the Corps: A Flood of Bad Projects," *The Washington Post*, May 14, 2006.

2. Such auctions have recently been proposed by the US secretary of transportation for airports in the New York region. See Mary E. Peters, "End Gridlock on the Runway," *The New York Times*, July 22, 2008.

3. The Office of Management and Budget does send notifications to all federal agencies urging the use of a 7 percent real discount rate for all capital projects, but it allows that other rates are appropriate under different circumstances.

4. See Lawrence Summers, "What We Can Do in This Dangerous Moment," *Financial Times*, June 29, 2008.

5. See the authors' "Measures to Avoid the Worst Recession in 30 Years," *Financial Times*, July 21, 2008.

2

Highways, Byways, and Railways: Transportation Infrastructure

A bicyclist rides near the Golden Gate Bridge on National Bike to Work Day, May 21, 2010, in San Francisco, California.

People drive on Highway 134 (Ventura Freeway) at the end of the evening rush hour in Glendale, California, on September 3, 2010, before the start of the three-day Labor Day holiday weekend.

Editor's Introduction

In 1817, after an intense lobbying campaign, New York governor DeWitt Clinton convinced the state legislature to allocate a total of $7 million towards the construction of a 363-mile canal connecting the Hudson River to Lake Erie. The canal took eight years of hard work to build, and during that time enthusiasm for the project waxed and waned. Skeptics mocked the governor's vision, dubbing the canal "Clinton's Folly" and "Clinton's Ditch." To a degree their doubts were understandable. Clinton's rhetoric tended toward the grandiose. The canal, he judged, "may prevent the dismemberment of the American Empire. As an organ of communication between the Hudson, the Mississippi, the St. Lawrence, the Great Lakes of the north and west and their tributary rivers, it will create the greatest inland trade ever witnessed." With the completion of the waterway, Clinton predicted, New York City "will, in the course of time, become the granary of the world, the emporium of commerce, the seat of manufactures, the focus of great moneyed operations and the concentrating point of vast, disposable, and accumulating capital, which will stimulate, enliven, extend, and reward the exertions of human labor and ingenuity, in all their processes and exhibitions."

A triumph of engineering, the Erie Canal was completed on October 26, 1825, and quickly transformed the nation. By effectively connecting the Atlantic Ocean with the Great Lakes, the canal opened up the West to settlement and development. Pioneers heading to the frontier no longer had to rely on costly and arduous overland routes. Additionally, as people and products traveled back and forth through the canal, Buffalo, Rochester, Syracuse, and other western New York cities along the waterway prospered. Commerce grew at a rapid pace, and New York City soon eclipsed Boston and Philadelphia as the nation's principal seaport and financial center. In hindsight, there had been nothing farfetched about Clinton's grand vision: his dream had proven prescient.

As the United States expanded over the continent in the decades that followed, other transportation innovations reshaped the nation. Railroads soon branched out across the landscape, connecting the country as never before. Before long, the airplane and automobile arrived on the scene. The selections in this chapter examine the history and present state of American transport infrastructure while also focusing on the themes and conflicts that may influence the future of our transit facilities. The first piece, "Transportation Infrastructure," excerpted from the America on the Move exhibit at the Smithsonian Institute's National Museum of American History, offers a quick survey of American transportation history, exploring the nation's early roads, the canal craze ushered in by the completion of the Erie Canal, the development of railways, the road-building frenzy that occurred following the emergence of the automobile, and the birth of air travel.

In "Transportation Infrastructure: Moving America," Robert McMahon considers the current condition of our transit networks and how President Obama's February 2009 stimulus package will impact them. In addition, he provides a primer on some of the major infrastructure initiatives of the 20th century, such as President Dwight D. Eisenhower's Interstate Highway System and certain New Deal projects implemented by President Franklin Delano Roosevelt. One issue McMahon contemplates is the future of high-speed rail, perhaps today's most important and divisive U.S. transit-related topic.

Ashley Halsey III discusses an alarming study issued by a bipartisan panel of transportation experts in the next article, "Failing U.S. Transportation System Will Imperil Prosperity, Report Finds." The panelists concluded, in Halsey's words, "the United States is saddled with a rapidly decaying and woefully underfunded transportation system that will undermine its status in the global economy unless Congress and the public embrace innovative reforms." Currently, most revenue for road maintenance and construction is generated through gasoline taxes, but during periods of economic hardship or elevated gas prices, people log fewer miles behind the wheel, which means less money for transport infrastructure. The study recommended that instead of taxing gasoline, the government ought to charge drivers a fee of two cents per mile. Yonah Freemark delves deeper into the funding question in his piece, "Cars, Highways, and the Poor," examining various proposals, from the vehicle-miles-traveled (VMT) tax to congestion pricing. Mark Laluan further explores the funding issue in the subsequent selection, "The U.S. Must Invest in Infrastructure." Laluan criticizes both political parties for not making transit facilities a budgetary priority and warns that the United States could leave itself at a competitive disadvantage if it does not invest in overhauling its transport grid.

Bruce Katz and Robert Puentes critique how transportation funds are distributed in the subsequent piece, "Clogged Arteries: America's Aging and Congested Road, Rail, and Air Networks Are Threatening Its Economic Health." They maintain that 100 metropolitan areas account for 75 percent of the country's economic production, but of 6,373 government-funded transportation projects, only half were focused on these vital regions. "In a post-agricultural, post-industrial, innovation-dependent economy, the roads to prosperity inevitably pass through a few essential cities," the authors observe. "We should make sure they're well maintained."

In "King of the Road," Alex Marshall points out what he sees as a disjunction in the transit infrastructure debate. According to Marshall, many conservative and libertarian groups support government spending on road and highway maintenance and construction but tend to oppose investment in rail projects. Such groups, Marshall contends, see "a highway as an expression of the free market and of American individualism, and a rail line as an example of government meddling and creeping socialism."

Given fears of global climate change, the political drawbacks associated with our dependence on foreign oil, and other concerns, many have argued that we need to shift away from the automobile as our primary mode of travel. In the last piece, "Shifting Gears: Breaking Our Addiction to Cars," David Carroll Cochran catalogs the steep costs of our love affair with the automobile, among them "[g]rowing traffic jams, pollution, the loss of prime farmland and open space to sprawl, weaker social ties, tacky strip-malls and endless big-box stores, the decline and isolation of the nation's urban cores, the social marginalization of those who don't drive—in particular, the very poor and old." In light of all these negative impacts, he calls on the United States to reassess its relationship with the car.

Transportation Infrastructure[*]

Excerpted from America on the Move
National Museum of American History, The Smithsonian Institute

It's hard to imagine America in 1800. The young country consisted of 16 states and just over 5 million citizens. The vast and impenetrable landscape made travel difficult and as a result people tended to live very local lives. But over the next hundred years, roads were built, canals dug, rivers improved, and rails laid, which allowed Americans to spread out and conquer the continent.

Road construction was one of the first improvements in American infrastructure. Major cities in the northeast were often connected by post roads, which at first were little more than dirt trails but later were improved with gravel or wooden planks. Travel on these roads was slow going—the trip from Boston to New York, for example, could take up to 3 days by stage coach.

In 1806, Congress allotted funds for the national road, the first federally funded road. It stretched from Cumberland, Maryland to, eventually, southern Illinois.

To reach points further west, hundreds of thousands of intrepid souls embarked on journeys from the banks of the Missouri. They headed out on horseback and in wagon trains, over routes like the Santa Fe and Oregon trails. The grueling trip all the way to the Pacific coast could take up to 8 long months.

Canals also helped link up the interior of the country. In 1825, the Erie Canal—the nation's most famous—opened for business. With it, food, goods and people could flow between New York City and the burgeoning west. Its success sparked a canal building boom throughout the eastern United States and elevated New York City to the nation's commercial center.

Engineering projects made navigation in coastal waters and along rivers safer and easier. Steamboat companies took advantage of these changes to move increasing numbers of passengers and cargo.

And although people and goods crossed the oceans at the beginning of the century, by its end, harbor improvements allowed more and larger ships, laden with immigrants and goods from Europe and Asia, to dock at U.S. ports.

[*] Reprinted by permission of the National Museum of American History, The Smithsonian Institute.

But nothing affected America's westward expansion like the growth of the railroad. It was fast and stupendous in scope: in 1840, there were 3,000 miles of track in the country; twenty years later, there were more than 30,000.

By 1869, enormous investment and spectacular engineering feats allowed the railroad to reach from coast to coast. After that first transcontinental link was made, hundreds of thousands of miles of rail were laid, consolidating the railroad's grip on the nation's long distance travel and trade.

By the turn of the century, the nation's growing network of canals, roads, waterways, and railroads had forged new links between people and places, and helped create the spectacular growth of 19th century America.

Between 1900 and 1950, the United States paved the way to its future. Local, state and federal dollars built millions of miles of roads, opening up new worlds to those who traveled along them.

You might think car and truck owners were the force behind good roads. But the first paved roads were a result of a very different lobby. Bicyclists wanted smooth streets for an easier ride. And health reformers fought for them because they were easier to clean than dirt or cobblestone—an important consideration when horses produced over one million pounds of manure a day in some cities. Later, farmers pushed for roads to get their goods to market and truckers lobbied for them between cities.

Cities took responsibility for their own roads, taxing adjacent property owners to fund improvements. States and the federal government took the lead in outlying areas where some roads made travel nearly impossible. In 1916, Congress realized the importance of good roads to the nation's economy and allocated millions to improve them. States matched the funds with money raised from gas taxes, setting off a highway building boom across the country.

The nation's first superhighway—the Pennsylvania Turnpike—opened in 1940. The four-lane toll highway set the standard for the future. There were no traffic lights, no intersections, no steep hills, no sharp curves and no speed limit.

The federal government also invested in air transport and waterways. Starting in 1904, the Army Corps of Engineers dug, blasted, and tore out jungles to build the Panama Canal. When it was completed ten years later, the canal knocked 8,000 miles off the trip from New York to San Francisco making trade and travel between the east coast and the west faster, easier, and cheaper. The government also spent billions of dollars dredging rivers and harbors and building levees throughout the nation to improve shipping. The U.S. Post Office and military built airports across the country and lucrative government mail contracts gave airlines the business they needed to become established.

By the end of World War II, the U.S. had a well-developed network of routes—air, rail, and road—that linked every state and connected the country as never before.

From the air, America's transportation system looks like a massive and intricate spider's web. Roadways reach up and around, railroad tracks sweep across the plains, large and small airports dot the land, rivers and canals snake across the country, and ports spread for miles along the coast. Fifty years ago, however, the transportation landscape was very different.

The modern era of roads didn't begin until 1956 when President Eisenhower signed into law the Federal Aid Highway Act. With it, he authorized the construction of 40,000 miles of limited access roads creating the interstate highway system. It was the biggest public works project this country has ever seen, and was lauded as one of the most important, as well.

But the interstate system also generated controversy. In cities, poor neighborhoods were often targeted for destruction, leading to protests. Still, most interstates were built as planned—and they saw use beyond their designers' wildest dreams.

As more people traveled by car and plane, railroad companies gave up their passenger service. They concentrated, instead, on hauling freight. Throughout the 20th century, more tonnage went by rail than any other way. In the 1960s, railroads began to haul goods in standardized steel containers that could easily be moved from train to truck to ship.

These containers made transport easier and cheaper. But they demanded a whole new kind of port. They needed large areas of flat land—and good connections to road and rail, to keep the cargo moving.

Airports also expanded. Cities invested billions in acres of hangars and terminals and miles of runways . . . they hired thousands of workers. All to get us where we wanted to go, when we wanted to get there.

Transportation Infrastructure[*]

Moving America

By Robert McMahon
Council on Foreign Relations, February 24, 2009

INTRODUCTION

Transportation experts view the call for dramatic federal government action in response to the economic crisis as an opportunity to overhaul the U.S. system of highways, bridges, railways, and mass transit. A series of sobering report cards from the American Society of Civil Engineers documents the inadequacy of this system. President Barack Obama took office pledging to act; his February 2009 stimulus package provides nearly $50 billion for transportation infrastructure. But many experts look beyond the stimulus and call for shifts in longer-term policy that will fundamentally alter the approach to planning and funding infrastructure and bolster U.S. competitiveness, quality of life, and security. In the past, the United States has revamped its transportation infrastructure to build canals, transcontinental railways, and a federal highway system, in each case helping usher in periods of economic growth.

A STATE OF DISREPAIR

A January 2009 report by the American Society of Civil Engineers on infrastructure, much of it involving the transportation sector, concluded: "all signs point to an infrastructure that is poorly maintained, unable to meet current and future demands, and in some cases, unsafe." It found that aviation, transit, and roads, already rated abysmal four years ago, had declined even further. Lost time from road congestion, the report estimated, was costing the economy more than $78 billion dollars a year while nearly half of U.S. households still had no access to bus or rail transit.

At the same time, national spending on infrastructure is often depicted as a faulty, wasteful process. Annual federal spending on transportation infrastructure in recent years has averaged more than $60 billion, and billions have been spent since 9/11 on aviation security. The Congressional Research Service cites Transportation Department data showing that the number of structurally deficient bridges was cut nearly in half between 1990 and 2007 due to federal spending. But 2006 Federal Highway Administration statistics also showed that more than 70,000 bridges, about 12 percent of the total, were structurally deficient. Among them was the I-35W bridge in Minneapolis that collapsed in August 2007, a mishap that killed thirteen people and spawned new debate about the focus and level of U.S. infrastructure spending.

There is a further homeland security dimension, says CFR Senior Fellow Stephen E. Flynn. He refers to the current state of U.S. infrastructure as the "soft underbelly" of the nation's security. "This is a core vulnerability for U.S. society," Flynn told a January 2009 CFR meeting. "It's very costly, after things fall apart, to try to put them back together again. And so, as I would forecast more generally in the twenty-first century, infrastructure is going to be [an] appealing target" to terrorists. Former Homeland Security Secretary Michael Chertoff has also expressed concern about the federal government's failure to make long-term infrastructure investments to overcome degradation of roads, bridges, dams, and other such "common goods."

Many experts say transportation infrastructure spending over the past several decades has failed to keep pace with the increasing burden absorbed by the country's roadways, bridges, and mass transit networks. The nonpartisan Congressional Budget Office shows that spending for infrastructure relative to gross domestic product (GDP) declined about 20 percent from 1959 to 2004. A number of experts also point to flaws in the manner of funding and planning U.S. infrastructure. Part of that involves abuses in the congressional earmarking process, epitomized by Alaska's so-called Bridge to Nowhere, a once-approved plan, later cancelled, for federal funding to build a $200 million bridge to a remote island.

Another serious problem is coordination between different forms of transportation, experts say. Congress aligns transportation funding with specific modes like highways, rail, and mass transit. It sought to improve coordination between these modes through legislation originating with the 1991 Intermodal Surface Transportation Efficiency Act. But the legislation did little to alter the congressional appropriations approach, the Government Accountability Office, a government watchdog body, found in a 2007 report. "As a result," the report says, "there is little assurance that projects, including intermodal projects—which could most efficiently meet the nation's mobility needs—will be selected and funded." The failure to achieve such coordination, note Brookings Institution experts Bruce Katz and Robert Puentes, leaves the United States as "one of the few industrialized countries that fails to link aviation, highways, freight rail, mass transit, and passenger rail networks."

STIMULUS AND TRANSPORTATION

During the country's greatest economic crisis, the Great Depression, President Franklin D. Roosevelt's New Deal was responsible for a series of iconic public works projects, notable among them the Golden Gate Bridge, the Hoover Dam, and the Tennessee Valley Authority rural electrification effort. No less important were "thousands of railroad grade crossings, parkways, trails in the national parks," writes historian Bruce Seely in the Wilson Quarterly. "Public-works relief funding from the federal government finally broke fiscal logjams."

The 2009 stimulus package is more modest, with its supporters stressing the need for speed in creating jobs as opposed to longer-term legacy projects, at least initially. The legislation includes roughly $48 billion in transportation infrastructure spending, including $29 billion for highway projects, $8.4 billion for public transportation, and $9.3 billion for inter-city rail, including up to $8 billion for high-speed rail service.

Many of the funds will be disbursed under normal formulas. State governments will receive funding for roads and bridges, and funds for public transit will go to local and regional agencies. The Obama administration and Democratic congressional leaders have stressed there are no earmarks in the bill; by contrast, the current federal surface transportation authorization bill that expires in September 2009 has more than 6,000 earmarks.

Still, the allotment for high-speed rail has piqued the interest of transportation experts. Most of the extra funding was inserted at the last moment by the Obama administration, signaling the new president's commitment to a form of transport in which the United States is seen as lagging far behind states like France, Germany, parts of coastal China, and Japan. Transportation Secretary Ray LaHood has sixty days to prepare a strategic plan for the funds and has the discretion to give "priority to projects that support the development of inter-city high-speed rail service," Obama's Chief of Staff Rahm Emanuel told Politico.

But some experts stress the need for realistic expectations about how such a system can function in the United States. "I'm not convinced high-speed rail is the answer to some people's prayers, because the geography of the United States is different than Europe," says William J. Mallett, a specialist in transportation policy at the nonpartisan Congressional Research Service. "It may make sense in certain places—the Northeast Corridor, California, Chicago. A high-speed rail network that covers the whole country is probably not feasible" due to the dispersed nature of the U.S. population, he says.

Yet overall, the stimulus spending directed at transportation will be helpful in starting to address some of the country's infrastructure shortcomings, says the Brookings Institution's Puentes. "From a lot of indicators, this does seem a step in the right direction," he said. "At the end of the day, it could have been a lot worse."

The exact apportionment of funds will be clear in the months ahead. In December 2008, the American Association of State Highway Transportation Officials listed more than 5,100 road and highway projects, $64 billion of which could get underway quickly, and the Association of Public Transit Officials points to more than $12 billion of "ready to go" transit projects.

Obama's stimulus package has aroused debate among economists about the usefulness of large public works projects to spur employment and economic growth. Overall, the 2009 stimulus package includes $120 billion for infrastructure and scientific research and more than $30 billion for infrastructure projects related to energy. Obama administration advisers estimate that construction projects alone could create about 675,000 jobs. Republicans in the House and Senate, nearly all of whom voted against the stimulus measure, have generally acknowledged the need for infrastructure improvements but have faulted the timeliness and targeting of the money in the package. Ronald D. Utt, a senior research fellow for the conservative Heritage Foundation, says it might have been wiser to stimulate employment in the battered financial sector than in major public works projects. "It is highly unlikely that any of these unemployed investment bankers, mortgage brokers, actuaries and bond counsel, and other skilled workers who make the financial system work would instead bang rivets into West Virginia bridges or hang drywall in Chicago schools," he wrote in a February 2009 web memo.

EISENHOWER'S EXAMPLE

State and local governments control an estimated 80 percent of infrastructure spending, giving the federal government more of a coordinating role. But there are precedents for the federal government asserting primacy in transportation infrastructure. The last major example was the 1956 legislation creating the Interstate Highway System, which led to the creation of 47,000 miles of highway and more than 55,000 bridges. President Dwight D. Eisenhower saw firsthand the parlous state of the nation's roads when he took part in a 1919 U.S. Army convoy from Washington D.C. to San Francisco. The journey took sixty-two days. As commander of allied forces in World War II, Eisenhower also took note of the efficiency of the German autobahns. As president, Eisenhower presented the highway plan as a national security imperative (the system is known as the Dwight D. Eisenhower National System of Interstate and Defense Highways).

Eisenhower "gave the rationale for the federal government to essentially put a national system in place where we had really a patchwork quilt of roads," CFR's Flynn told the January 2009 meeting. The effort was funded by a highway trust fund, which received funding primarily through a federal tax on gasoline.

Flynn says security can again be used as a rationale to spur a more coherent, nationally oriented approach to infrastructure. Another panelist at the January

2009 CFR meeting, Everett Ehrlich, a former undersecretary of commerce, said a standardized review of all infrastructure efforts at one federal "checkpoint" would help planners address everything from smoother transit to related concerns such as homeland security and energy efficiency at one time.

BEYOND THE STIMULUS

As a presidential candidate, Obama expressed support for a National Infrastructure Reinvestment Bank that would invest $60 billion over ten years for improvements to "maximize our safety and homeland security." The notion of a national bank for infrastructure has gained a number of prominent adherents in recent years, including governors and mayors. A 2007 report commissioned by the Center for Strategic and International Studies called for a bank that would draw together U.S. infrastructure proposals and collect a portfolio of investments to fund them. As described by Ehrlich and a co-chair of the CSIS commission that produced that think tank's report, Felix G. Rohatyn, the National Infrastructure Bank would preserve the system in which state and local governments propose major infrastructure investments, but "would change dramatically the way priorities are set and projects funded" by eliminating the separate programs for highways, airports, and mass transit.

In a similar vein, Puentes of the Brookings Institution calls for establishing a Strategic Transportation Investments Commission. It would focus on maintaining the interstate highway system, developing a national intermodal freight plan, and creating a national plan for passenger travel among metropolitan areas. The political viability of these and numerous other proposals for infrastructure overhaul, including increased use of public-private partnerships, will get a supreme test in the coming months, as Congress is due to reauthorize the 2005 Surface Transportation Extension Act, which expires on September 30, 2009.

Failing U.S. Transportation System Will Imperil Prosperity, Report Finds[*]

By Ashley Halsey III
The Washington Post, October 4, 2010

The United States is saddled with a rapidly decaying and woefully underfunded transportation system that will undermine its status in the global economy unless Congress and the public embrace innovative reforms, a bipartisan panel of experts concludes in a report released Monday.

U.S. investment in preservation and development of transportation infrastructure lags so far behind that of China, Russia and European nations that it will lead to "a steady erosion of the social and economic foundations for American prosperity in the long run."

That is a central conclusion in a report issued on behalf of about 80 transportation experts who met for three days in September 2009 at the University of Virginia. Few of their conclusions were groundbreaking, but the weight of their credentials lends gravity to their findings.

Co-chaired by two former secretaries of transportation—Norman Y. Mineta and Samuel K. Skinner—the group estimated that an additional $134 billion to $262 billion must be spent per year through 2035 to rebuild and improve roads, rail systems and air transportation.

"We're going to have bridges collapse. We're going to have earthquakes. We need somebody to grab the issue and run with it, whether it be in Congress or the White House," Mineta said Monday during a news conference at the Rayburn House Office Building.

The key to salvation is developing new long-term funding sources to replace the waning revenue from federal and state gas taxes that largely paid for the construction and expansion of the highway system in the 1950s and 1960s, the report said.

"Infrastructure is important, but it's not getting the face time with the American people," Skinner said. "We've got to look at this as an investment, not an expense."

A major increase in the federal gas tax, which has remained unchanged since it was bumped to 18.4 cents per gallon in 1993, might be the most politically palatable way to boost revenue in the short term, the report said, but over the long run, Americans should expect to pay for each mile they drive.

"A fee of just one penny per mile would equal the revenue currently collected by the fuel tax; a fee of two cents per mile would generate the revenue necessary to support an appropriate level of investment over the long term," the report said.

Fuel tax revenue, including state taxes that range from 8 cents in Alaska to 46.6 cents in California, have declined as fuel efficiency has increased. President Obama mandated that new cars get 35.5 miles on average per gallon by 2016, and government officials said last week that they are considering raising the average to 62 miles per gallon by 2025.

Facing midterm elections in November, Congress has lacked the will to tackle transportation funding. Efforts to advance a new six-year federal transportation plan stalled on Capitol Hill after the previous one expired last year.

If Congress were to do the report's bidding, the task would be far broader in scope than simply coming up with trillions of dollars in long-term funding to rebuild a 50-year-old highway system.

The experts also advocated the adoption of a distinct capital spending plan for transportation, empowering state and local governments with authority to make choices now dictated from the federal level, continued development of high-speed rail systems better integrated with freight rail transportation, and expansion of intermodal policies rather than reliance on highways alone to move goods and people.

But Mineta noted that 42 days after an eight-lane bridge collapsed into the Mississippi River in Minneapolis in 2007, a survey found that 53 percent of respondents opposed an emergency gas tax increase to pay for infrastructure repairs.

"The shelf life of a tragedy like [I-35W] was 42 days," he said. Thirteen people died in the collapse and more than 100 were injured.

The report emphasized that federal policy should be crafted to address congestion by providing incentives that encourage land use that reduces single-occupant commutes and promotes "liveable communities."

"Creating communities conducive to walking and alternate modes of transportation . . . should be an important goal of transportation policy at all levels of government," the report said.

It also encouraged expansion of innovative public-private partnerships to further transportation goals, citing the high-occupancy toll lane project in Northern Virginia as an example.

"The one option that's not in this report is throwing up our hands," said Jeff Shane, a former Transportation Department official and a member of the panel. "That seems to be the option that Congress chooses."

Cars, Highways, and the Poor[*]

By Yonah Freemark
Dissent, Winter 2010

The infusion of money into highways through gasoline taxes and the suburban exodus of the middle class and their adoption of the automobile as the primary mode of transportation profoundly altered the nation's landscape over the past fifty years. The trend seemed inevitable: America would be a nation of drivers residing in unwalkable, monofunctional neighborhoods far from city centers.

When gas hit four dollars a gallon in 2008, however, something exceptional happened: for the first time in history, the number of miles driven in passenger vehicles dropped. Though fuel prices have declined, the reduction in gas consumption is a long-term trend. Hybrids and small cars are replacing gas-guzzlers; eventually, electric vehicles that use no taxable fuel at all may become the norm.

Although progressive transportation activists hail this decline, they must face the challenge of advocating policies that do not adversely affect the poor, who now live mainly outside of inner cities and rely on cars for access to jobs, food, schools, and services. The first quandary is how to fund desperately needed road improvements while simultaneously pushing for changes that will make mass transit more feasible in suburban areas.

For decades, U.S. transportation policy prioritized the completion of the Interstate Highway System, whose multi-lane freeways now span out across the country like a spider's web. The program was easy to understand: drivers paid taxes on their gasoline consumption, Washington replenished the Highway Trust Fund, and revenues were redistributed to the states to construct new roads. The system was self-reinforcing: the more people in cars, the more concrete could be laid. At last count, the network was more than 45,000 miles long.

The current decline in car use spells disaster for the Fund. Having already obligated billions to states ready to build new roads and transit systems, Congress in 2008 had no choice but to dip into the general treasury and authorize an $8 billion infusion of income tax revenues to fill the Fund's emptied coffers. In 2009, facing the same difficulties, the government repeated the action—*twice*. But Congress has yet to find a long-term way to replenish the Fund. Where will the money come from?

There is no easy answer. In the short term, Congress will likely continue to rely on fill-up money from the general treasury. But the failure of the fuel tax to cover the costs of the transportation system suggests a more general failure: the current system does not meet the needs of a modern America in which rich and poor alike increasingly live in the suburbs.

For fifty years, the idea that a user fee of some sort should lie at the heart of the funding for any transportation project reigned supreme. As a result, this recent switch to income-tax-derived funds is a paradigm shift—and a great opportunity. We can now imagine revenues being raised in a way that would encourage the development of a more equitable and environmentally conscious society less reliant on the automobile.

Washington is mired in debt, but cutting funding for transportation is not an option; the United States spends only 2 percent of GDP on infrastructure compared to 5 percent for most European countries and 10 percent for China. The difference shows in the deterioration of our roads and bridges. A recent study by the American Society of Civil Engineers projects a five-year shortfall of more than $750 billion in the expenditures necessary to maintain today's ground transportation system.

In the early 1990s, Congress twice increased fuel charges to make up gaps in the system, and that solution beckons again. "In the short-term, the gas tax is the only realistic option," Robert Puentes of the Brookings Institution told me, citing its current use and the ease with which it could be raised. But most of the political world is against it. In his June 2009 Senate testimony. Secretary of Transportation Ray LaHood said that the White House would "oppose a gas tax increase during this recessionary period," but provided no clues as to how he would fund his agency.

Yet alternatives are just as difficult to propose during an economic downturn. One option is the vehicle-miles-traveled (VMT) tax, which would charge drivers for their use of the roadways. Key Democratic lawmakers have blasted it as a violation of personal privacy because it would require the installation of satellite-linked GPS units or the imposition of mandatory odometer checks. Another possibility is widespread tolling of the kind already seen along highways in the Northeast, but it is considered politically suicidal to charge people to drive on roads that were once free. Tolls also have the hard-to-manage tendency to increase traffic on other roads nearby. Neither solution would encourage the purchase of fuel-efficient vehicles, in contrast to the gas tax.

Any of these options would raise a political firestorm, but the user fee approach—whether based on gas consumption or road use—remains most attractive because it has garnered support on both ends of the ideological spectrum. Conservatives defend highway spending because it "pays for itself." Progressives like the ecological advantages of limiting driving by increasing its cost. Raising user fees would reduce the number of cars on the road. New revenue could be used to expand public transit, one of the major objectives of the Left's transportation policy.

"The U.S. really needs to learn more from the best global experience," said Michael Repogle of the Environmental Defense Fund in an interview, citing Singapore as a model for American policy. That country's use of an electronic toll on cars entering downtown has allowed it to reduce traffic congestion significantly since 1975, even as citizens have "increased their motor vehicle ownership by three times." The city-state has used the funds to expand public transportation to handle 66 percent of intra-island travel, compared with only 40 percent thirty years ago.

For years, New York City has considered implementing a similar congestion charge system that would assess a fee on car owners entering or driving in the central business district south of Manhattan's 60th Street. Despite criticism that the fee would hurt the poor, middle- and lower-income commuters would be minimally affected because they already have access to an extensive network of cheap bus and rail connections that would only be improved with the help of new funding.

This lesson can be generalized to other cities—rare as they may be—that have a strong public transit system. They are the only places whose urban poor would be the least affected by any reform that emphasizes higher automobile fees. Progressive proponents of user fees, however, overlook a massive demographic shift: the poor have moved out of cities.

In 2006, for the first time, more of the nation's impoverished families lived in the suburbs than in central cities—not to mention the millions living in sprawling municipalities whose form replicates suburbia in all but name. These people, who lack access to alternative modes of transportation, would be the most penalized by any approach that advocates user fees. Ironically, the federal transportation system's emphasis on highway construction has produced a car-dependent class whose members cannot afford any hike in commuting costs.

A case in point is Phoenix, Arizona, where sprawling suburbs and a relative dearth of bus and train service force nearly 90 percent of commuters to use their cars—this despite the fact that 16 percent of that city's inhabitants live in poverty. In contrast, only 33 percent of New York's population drives to work. In worst-case Arlington, Texas, population 400,000, there is no bus service at all, despite the city's 10 percent poverty rate.

Places like Phoenix and Arlington are the rule rather than the exception. If it was once accurate to stereotype American poverty as an inner-city phenomenon, it would be completely unacceptable to do so today. National housing and transportation policy has for decades favored extra-urban development, sponsoring investment in highways and giving incentives to homeowners. Those choices have encouraged sprawl; it is no surprise that the poor, looking for new opportunity, have followed the pioneering middle classes that first took advantage of life in the suburbs.

Attractive as they may be to families hoping to escape the desolation of inner-city ghettos, however, suburban landscapes of strip malls and cheap single-family homes are hostile to pedestrians and transit users. Dwellings are out of walking range of offices, retail outlets, and parks, making the use of cars obligatory. Unsurprisingly, public transportation in suburban settings often suffers from low ridership, because it is difficult to make it convenient or accessible in places designed for the car.

Although driving is expensive, millions of poor families have no alternative. Almost 30 percent of the nation's households have annual incomes lower than $25,000, yet more than 90 percent of Americans own a motor vehicle, at an average expense of $8,000 per year. Any increase in user fees designed to reduce car travel would fall hardest on the budgets of the working class, whose freedom of movement is already limited significantly because of the high cost of transportation.

How, then, to address the financing of the transportation system without reducing the mobility of the car-dependent poor? Some defenders of mass tolling or a VMT fee argue that Congress could provide rebates for the nation's poorest families. Experience suggests, however, that the desire for funds will supersede any urge to assist the poor. Existing toll roads—most of them in "progressive" northern states like New Jersey and Pennsylvania—provide no discounts based on income. If these states cannot address this problem, how can we count on a considerably more conservative Congress to do so?

Many progressives argue that user-fee revenues should be redirected to alternative transportation projects in poorly served areas. But could we transform suburbs into places where people have access to public transit and use it? Any significant transformation seems unlikely for one reason: suburban population densities are simply too low to support convenient transit networks, which need a large customer base to fill buses and trains and limit subsidies. The average neighborhood of single-family homes has a population density ranging from 2,000 to 5,000 people per square mile. By comparison, in Santa Monica, California, which has a density of 10,000 people per square mile and high-quality bus service, nearly 80 percent of people drive to work every day; places like New York, with extensive transit use, have much higher concentrations of population. Even a doubling or quadrupling of existing suburban densities and a corresponding increase in transit services would likely persuade only about 10 percent of the population to abandon their vehicles. Although a higher user-fee approach might force some people onto buses and trains, the vast majority would still find their automobiles more convenient for everyday use.

Redesigning the suburbs would take years and lots of money. New or improved public transportation facilities cannot be developed *now* with funds to be earned in the *future*. What happens in the meantime? Strong, well-funded programs that invest in infrastructure are imperative if the goals progressives espouse are to be realized.

Extending the use of general funds to finance transportation is a good place to start. The income tax source is progressive, as opposed to the user-fee system, which assumes that the poor should pay just as much as the rich. Progressive taxation provides the federal money for education and affordable housing, and it is the principle underlying expenditures on national defense.

The ability to move freely, like the need for housing and health care, is a basic necessity for every member of a modern society. Framing it as an essential public service could reduce resistance to using the income tax to pay for roads and transit.

Yet the user-fee model is so ingrained in the common understanding of how transportation systems should work that other possibilities are not easy to accept. "It's not the solution, only because people fear that having to haggle every year for a piece of the general fund means you can never be certain whether there's money or not," David Goldberg, communications director for Smart Growth America, a transportation and planning advocacy group, told me. "Relying on it exclusively is very unnerving for people in the transportation field." However, there is no technical reason why a portion of overall income tax revenues could not be regularly transferred to the Highway Trust Fund and guaranteed over a period of several years.

Because the system is supposed to pay for itself, it is difficult to argue against continuing to appropriate fuel tax revenues back into highways. As long as drivers continue to be the transportation system's primary funders, it is politically onerous to divert money away from roads and toward transit. Car use is thus reinforced. To make matters worse, the system breaks down when the number of drivers declines. "As our population grows, we're going to drive less per capita," says Goldberg, "which means less money into the transportation system."

Income tax-sourced revenue would address both problems, detaching the system from the tether that binds it to roads projects and breaking from the self-support concept. We won't see any immediate reduction in driving as a consequence of switching to a non-user-fee system, but we will ensure universal access to transportation even as we expand the potential for government to invest in alternative modes and slowly prepare the country for a less automobile-dependent future.

But changes in the funding mechanism won't be enough to deal with environmental concerns. Although cheap gas and free highways have produced the sprawl that so frustrates advocates of alternative transportation, too many other policies relating to land use and transit provision add to the problem. Laws that limit the construction of mixed-use neighborhoods and instead promote malls and culs-de-sac should be changed. Instead of sending billions of dollars to states to build new, mostly unnecessary, highways every year, Washington should encourage the construction of denser communities that foster public transit. Strip malls should be converted into walkable town centers. Local bus services should be ramped up.

Only *after* low-income families have ready access to a full range of living choices supported by alternative modes of transportation could fees on car use be increased without hurting those who are already suffering economically. Progressives should not find themselves sacrificing the mobility of the poor today to pay for transportation improvements that won't provide convenient alternatives to the car for years or even decades to come.

YONAH FREEMARK *is a freelance journalist based in Paris. He writes on transportation and urbanism and blogs at thetransportpolitic.com.*

The U.S. Must Invest in Infrastructure*

By Mark Laluan
The Pioneer Online, April 8, 2011

Between increasing vitriolic debates ranging from national defense to health reform, our elected officials have remained silent regarding improvements to America's transportation infrastructure.

America's National Highway System had its last major expansion undertaken under the Eisenhower Administration of the 1950s. While Americans have continued to rely on this increasingly dilapidated system of roads, bridges and highways, across the world states such as China have embarked upon full overhauls of their transportation systems.

China's National Trunk Highway System recently surpassed the American automobile based network with over 74,000 km of roads laid down. In addition, China has nearly quadrupled its railway network since the founding of the People's Republic in 1949 to over 91,000 km of rail lines.

These transportation networks are the arteries that supply the raw materials and transport the finished goods that fuel China's growth. Whereas China has worked to expand her transportation network, America remains content to sit on an increasingly hazardous network.

In 2007, the collapse of the I-35W Mississippi River Bridge in Minnesota serves as a chilling reminder of what has become willing neglect of America's transportation system. A bridge which was declared "structurally deficient" by the U.S. Department of Transportation in 2005 was passed over for replacement.

When the bridge collapsed under the weight of rush hour traffic, refurbishment focusing on replacing guard rails and pavement was underway. Vital retrofitting to the bridge's support structure would not be addressed in these repairs.

The result of such lackadaisical maintenance was a tragedy which claimed the lives of thirteen individuals and put an entire bus full of school children at risk of grave injury.

---

* Article by Mark Laluan originally published in *The New Pioneer*. Reprinted with permission of the author. All rights reserved.

The I-35W Mississippi River Bridge is symptomatic of an entire transportation system woefully neglected by our public officials. Democrats would rather spend trillions exclusively on social welfare programs such as socialized medicine, and Republicans have continued to harp almost exclusively on the need to decrease spending in all areas including infrastructure spending.

Thus both major parties are unable to present a cohesive program to rebuild America's roads, bridges, railways and airports. These vital links that enable the free flow of goods and services from one end of the country to the other impact the quality of life of all Americans.

We suggest that local, state and federal officials make transportation reform an honest priority. New forays into the Middle East—such as intervention in Libya—or costly social welfare programs—such as socialized medicine—have served to divide our nation's focus at a time when rebuilding our transportation infrastructure is crucial for the revival of our flagging economy.

We believe that while a measured response to foreign crises and the plight of medically uninsured is necessary, we also believe that it is crucial to ensure our transportation network functions to benefit economic expansion.

America depends on tax revenues from businesses as well as individuals to pay for national defense as well as social programs. Without an efficient transportation network, there will be no effective method to control the cost of living by shipping goods such as food to areas of the country where they are most needed.

Nor will America be in any shape to provide an appealing place for business to grow, develop and mature without a transportation network to ship goods upon.

More railways, roads, bridges and airports translates into more capacity for rapid transportation, which in the long run means lower fuel costs. This translates into real savings for the average consumer as businesses pass on the savings in overhead expenses to the consumer.

America is not awash with cash as pundit Michael Moore would have one believe. Our wealth and our capital are being tied up in overseas military adventures, social welfare programs and pork barrel spending, not in the hands of American citizens, rich or poor.

Reigning in such largess is the one step towards kick starting American economy. The second and equally crucial step would be to invest in a new transportation infrastructure, which can safely and effectively serve America's economic needs well into the future.

Clogged Arteries[*]

America's Aging and Congested Road, Rail, and Air Networks Are Threatening Its Economic Health

By Bruce Katz and Robert Puentes
Atlantic Monthly, March 2008

Transportation spending is spread around the United States like peanut butter, and while it's spread pretty thick—nearly $50 billion last year in federal dollars for surface transportation alone—the places that are most critical to the country's economic competitiveness don't get what they need. The nation's 100 largest metropolitan regions generate 75 percent of its economic output. They also handle 75 percent of its foreign sea cargo, 79 percent of its air cargo, and 92 percent of its air-passenger traffic. Yet of the 6,373 earmarked projects that dominate the current federal transportation law, only half are targeted at these metro areas.

In the past, strategic investments in the nation's connective tissue—to develop railroads in the 19th century and the highway system in the 20th—turbocharged growth and transformed the country. But more recently, America's transportation infrastructure has not kept pace with the growth and evolution of the economy. As earmarks have proliferated, the government's infrastructure investment has lost focus. A recent academic study shows that public investment in transportation in the 1970s generated a return approaching 20 percent, mostly in the form of higher productivity. Investments in the 1980s generated only a 5 percent return; in the 1990s, the return was just 1 percent.

[…] In most major metro areas, [traffic] is steadily worsening. The cost of congestion, including added freight cost and lost productivity for commuters, reached $78 billion in 2005. Half of that occurred in just 10 metro areas.

America's biggest and most productive metro regions gather and strengthen the assets that drive the country's prosperity—innovative firms, highly productive and creative workers, institutions of advanced research. And the attributes of some cities are not easily replicated elsewhere in the U.S. The most highly skilled financial professionals, for instance, do not choose between New York and Phoenix. They choose between New York and London—or Shanghai. While many factors affect that choice, over time, the accretion of delays and travel hassles can sap cities of their vigor and appeal. Arriving at Shanghai's modern Pudong airport, you can hop aboard a maglev train that gets you downtown in eight minutes, at speeds approaching 300 miles an hour. When you land at JFK, on the other hand, you'll have to take a train to Queens, walk over an indoor bridge, and then transfer to the antiquated Long Island Rail Road; from there, downtown Manhattan is another 35 minutes away.

To power our metropolitan engines, we need to make big, well-targeted investments that improve transportation within and around them. Above all else, that means taking a less egalitarian approach to our infrastructure: there is little justification for making small improvements all over the place.

In a post-agricultural, postindustrial, innovation-dependent economy, the roads to prosperity inevitably pass through a few essential cities. We should make sure they're well maintained.

BRUCE KATZ *is the director of the Brooking Institution's Metropolitan Policy Program.*
ROBERT PUENTES *is a fellow in the Metropolitan Policy Program.*

King of the Road[*]

By Alex Marshall
Governing, April 2008

The ancient Romans had a saying: To make a road straight, you need to make someone's neck crooked.

This chilling refrain is a vivid summing up of an obvious fact: Building a road is a manifestation of power, particularly state power. Carving a road across multiple jurisdictions and property lines—not to mention varying terrain—can be done only by an institution that can override the wishes of any one individual.

This was true in the days of the Roman Empire, when mighty roads were built so well that many of them still exist. And it's true today. In the exercise of that authority, local, state and federal governments spent more than $150 billion on roads in 2005, according to the most recent federal Highway Statistics report. That's comparable to what we spend annually on waging war in Iraq.

Given all this, I find it exceedingly strange that a group of conservative and libertarian-oriented think tanks—groups that argue for less government—have embraced highways and roads as a solution to traffic congestion and a general boon to living. In the same breath, they usually attack mass-transit spending, particularly on trains. They seem to see a highway as an expression of the free market and of American individualism, and a rail line as an example of government meddling and creeping socialism.

Among the most active of these groups is the Reason Foundation, a self-described libertarian nonprofit organization with a $7 million budget that has its own transportation wing. Some typical highway-oriented papers on Reason's Web site include "How to Build Our Way Out of Congestion" and "Private Tollways: How States Can Leverage Federal Highway Funds." Rail transit is taken on in papers with titles such as "Myths of Light Rail Transit," and "Rethinking Transit 'Dollars & Sense': Unearthing the True Cost of Public Transit." I didn't see any papers about unearthing the true cost of our public highway network.

Many of the authors of these studies are a rotating cast of writers who pop up again and again, including Randal O'Toole and Wendell Cox. They "extol the autonomy made possible by automobiles" wrote fellow libertarian and New York Times columnist John Tierney in a 2004 article on the subject. Tierney calls them, including himself, "the autonomists." That is, libertarians who have embraced highway spending, although they focus more on the individually-bought car, not the government-built road it requires.

Reason Foundation's founder and former president, Robert Poole, leads the group's Transportation Studies wing, and it's clear he has a special love of the subject. He has authored many studies himself, and he puts out the Surface Transportation Innovations newsletter. In an interview, I ask him to square Reason Foundation's support for roads with its general dislike of government involvement.

"I'd never thought about it that way," he says. Poole insists Reason doesn't want to eliminate government from transportation. "We aren't going to have competing companies putting roads in where they like, and letting the chips fall where they may. We aren't anarchists."

But the organization does have a general premise, which, Poole says, "is that transportation infrastructure would work better if it were market-driven. Where it's possible, that infrastructure should be run in a business-like manner with users paying full cost." All of this sounds good but is essentially incorrect. Transportation is like education: It works best through heavy general funding that pays off down the road in a community's or nation's overall prosperity. Our national road system would never have been built if every street were required to pay for itself.

Governments at every level have put in several trillion dollars' worth of roads over the past century. This system, open to all with a car, has created our automobile-based landscape of suburbs, single-family homes, office parks, mega churches and shopping malls. Love it or hate it, it is the product of massive government spending. As others have pointed out, the national road system is one of the most successful examples of pure socialism to be found: a comprehensive public system, well-used, almost entirely paid for with tax dollars.

Some of Reason's transportation ideas, such as truck-only toll lanes and congestion pricing, are worth considering. But the systematic bias in favor of roads and against mass transit makes the foundation's work suspect. City and state officials, who are frequently confronted with its studies, should view the work skeptically.

Shifting Gears[*]

Breaking Our Addiction to Cars

By David Carroll Cochran
Commonweal, March 13, 2009

It's no secret that we Americans have spent the past half-century building our communities and our lives around the car, which gives us the freedom to go anywhere we want, whenever we want. Cars give us other things too: the safety and room to spread out in the suburbs; the social status signaled by the kind of car, truck, or SUV we drive. These advantages and aspirations have given rise to the car-dominated American landscape we see around us today.

It's also no secret that this new landscape comes with significant costs. Growing traffic jams, pollution, the loss of prime farmland and open space to sprawl, weaker social ties, tacky strip-malls and endless big-box stores, the decline and isolation of the nation's urban cores, the social marginalization of those who don't drive—in particular, the very poor and the old.

As these costs have grown, so too has an awareness that our current car-centered habits are neither desirable nor sustainable, and that the time has come for a real shift in priorities. New political and economic circumstances make such a shift possible. Dramatic swings in gas prices over the past year and the probability of rising prices in the long-term future, an unprecedented crisis in the American automobile industry, and the prospect of massive infrastructure spending as part of government stimulus efforts—together these things provide an opportunity for fundamental changes in how we build, and move around in, our communities.

Our dependence on cars is not just an economic problem; it is also a moral problem. Transportation or zoning policies are not usually considered "values" issues, but they are. The values underlying our society's addiction to the car should trouble anyone who believes in strong, vibrant, and inclusive communities, in care for the most vulnerable members of society, and in environmental stewardship. The culture of the car and the landscape it fosters promote the private realm of co-

cooned comfort—the personal automobile, the large suburban home, the fenced backyard, the gated development. It neglects and debases the public realm, turning it over to highways, parking lots, and endless stores whose monotonous drabness is covered by garish signs or fake façades.

Some insist that transportation is simply a matter of individual choices: if people drive a lot, that's because they prefer to drive. And, of course, there is some truth to this. The attraction of suburban living and the convenience and flexibility of cars is undeniable, which is why suburbs and cars are not about to disappear. What this explanation misses, however, is the way our individual choices depend on the range of options we face and the incentives or disincentives attached to those options. The context in which we make choices is strongly shaped by public policies. The rise of the car and patterns of residential development were and continue to be driven by state, local, and federal policies that intentionally promoted driving and the kind of suburbs that driving makes possible. Such policies were not inevitable, and they are not irreversible. Different policies could still restore what has been damaged or lost.

Cars are by far the most subsidized form of transportation in the country. In addition to the public resources we devote to securing a fuel supply and to registering, licensing, and policing, cars aren't usable without an enormous infrastructure of government-built roads and government-mandated parking. If most roads and parking seem "free" to drivers, it is only because public policy makes sure they will be there at enormous public expense. In the United States cars are a "sunk cost."

Once we buy a car, the cost of each use is relatively small. And public policy has insured that the alternatives remain expensive, unattractive, or unavailable. Zoning requirements that separate and spread out most of the places we want to go make walking (the oldest, cheapest, and most reliable form of transportation) unrealistic. Same goes for biking. Except for a few notable exceptions, public transportation has been neglected or abandoned. By the 1950s, an alliance of car, oil, and tire companies had successfully (and illegally) bought and closed most of the nation's trolley lines. Since then, political influence has insured that government support for cars massively exceeds that for trains, light rail, and bus lines. This has left public transportation far less convenient and reliable than it is in many other countries. And yet those who do use public transportation must pay more per ride than most drivers pay for each use of their cars.

It is time to move toward a much more mixed transportation infrastructure, one where the car plays an important but much smaller role. This means shifting subsidies away from roads and parking, and adopting policies that make drivers pay a higher percentage of the actual cost of driving. Such measures might include higher gasoline taxes, tying automobile registration rates to annual odometer readings, reducing the number and raising the cost for parking spaces, and charging fees for driving in certain urban zones, as London now does—and as the mayor of New York City recently tried to do. As the price people pay to drive comes closer to the true cost of driving, more people will turn to public transportation, which must, in turn, receive greater public support. Many countries around the world, as well as some communities in the United States such as Portland, Oregon, have developed well-integrated public transportation systems, combining bike and walking paths, trains, light rail, and buses. New initiatives such as dedicated bus and bike lanes, fare-free zones, and shared city-bike programs deserve much more support across the country.

For such shifts in transportation policy to work most effectively, we also need to reconsider the physical layout of our communities. In many parts of the country, zoning laws still mandate sprawl. They require new construction to be separated by function (homes and apartment complexes here, commercial strips over there) and linked by feeder roads. Because of such arrangements, populations are too dispersed to support effective mass transit. More traditional forms of city planning wove homes, rental units, shops, parks, restaurants, churches, schools, and civic buildings into an integrated fabric, on a scale suitable for walking or busing. Such planning usually paid as much attention to shared spaces—the design and layout of streets, plazas, and prominent public structures—as to the inside of buildings. This is how the most attractive and humane towns and cities in the country were built, in stark contrast to the suburban sprawl and monotony that have characterized the last half-century.

As the problems of sprawl-style zoning have become more apparent, architects and planners have begun turning back to older patterns for inspiration. Public policies have begun shifting as well, but not quickly enough. A return to mixed-use and higher-density zoning is urgent. Of course, many towns, city neighborhoods, and inner-ring suburbs survive as models of this kind of zoning. This is why planning should favor the protection and rehabilitation of existing infrastructures. As for new construction, developers should be encouraged to fill in the empty spots downtown before they build on the periphery of cities. And while outward growth is inevitable in many places, it too can aim for walkable mixed-use neighborhoods close to public-transportation corridors.

The car is not about to go away. Neither will the need to account for its use when we plan new construction. But making the car such a dominant part of our lives and our landscapes has done much harm to both. The time has come for a change in cultural priorities and public policies. As a country we are currently facing other very difficult political challenges. But some of these—notably in the interrelated areas of economic stimulus, infrastructure, transportation, energy, and environmental protection—present new opportunities. They give us a chance to develop much more sensible long-term approaches to city planning and public transportation. Together, old environmental concerns and new economic conditions may finally break our addiction to cars.

DAVID CARROLL COCHRAN *teaches politics and directs the Kucera Center for Catholic Intellectual and Spiritual Life at Loras College in Dubuque, Iowa.*

3

The Train Debate:
Can Rail Revolutionize American Transport?

An Amtrak Acela train at Philadelphia's 30th Street Station on National Train Day, May 8, 2010.

Chicago Transit Authority (CTA) workers survey the wreckage of a CTA elevated train after it derailed south of downtown on May 28, 2008, in Chicago, Illinois. Only minor injuries were initially reported from the accident.

Editor's Introduction

Before the automobile came on the scene, just about every U.S. city with a population of more than 10,000 had streetcar or trolley service. As more and more Americans started to drive, these local transit systems were neglected. Eventually, most were either replaced by buses or closed altogether. Some conspiracy theorists even claim that General Motors (GM) and other auto manufacturers contributed to their decline, buying up local railways through front companies in order to dismantle them and make the automobile the only form of municipal transit available to the public.

Just as the car supplanted local rail service, the airplane largely replaced long-distance train travel. Whereas a trip from New York City to Los Angeles involves several days on a train, by plane it is a mere six hours. Still, while few expect the railroads to replace the airlines when it comes to cross-country travel, many believe that the train is poised for a revival as a mode of both local and regional transit. Indeed, some see in increased rail use a possible panacea to the sprawl, pollution, gridlock, and other excesses associated with the automobile. Articles in this section explore the potential railway renaissance and whether the enthusiasm of train advocates is warranted or if the anticipated benefits are exaggerated.

Across East Asia and much of Western Europe, high-speed rail (HSR) has emerged as a prevailing form of mass transit. Traveling at speeds in excess of 100 miles per hour, trains pick up and deposit passengers with startling efficiency and without the headaches Americans often associate with the daily commute. Yet in the United States, HSR—and trains in general—take a backseat to the highway. In "Bullet Trains for America?" Mark Reutter considers why this is so while also examining the various HSR projects currently underway. Peter A. Harkness focuses on commuter trains in addition to HSR in the next piece, "All Aboard? Bullet Trains May Be Sidetracked, But Not Commuter Rail," and considers the fate of President Barack Obama's proposed railway projects in light of the Republican victories in the 2010 midterm congressional elections.

While such cities as New York, Boston, Chicago, and Philadelphia have had municipal rail service for over 100 years, other locales have much newer systems. Atlanta, Washington, D.C., and the San Francisco Bay Area all constructed their subway systems in the 1970s. Since the 1980s, however, ambitious city planners have moved away from subways in addressing their transport needs, as Zach Patton reports in "Leaping into Light Rail: Cities in the South and West Are Writing a New Chapter in the History of Transit." The opening of a light-rail system in San Diego in the early 1980s marked a sea change in municipal transit circles, as subway construction fell out of favor and less costly light rail took up its place. In recent years, light-rail lines have launched in Charlotte, Denver, Houston, and Phoenix, among other cities, and Patton predicts that more municipalities may soon be climbing aboard.

In the subsequent selection, "Accelerating Amtrak: Will Superfast Train Service in the Northeast Ever Happen?" Alex Marshall discusses the potential for HSR in the Northeast Corridor, the densely populated coastal pocket stretching from Washington, D.C., to Boston. One reason HSR has failed to catch on in the United States the way it has in Western Europe and elsewhere is that our population tends to be more far flung, the distance between our cities greater. In the Northeast Corridor, however, the settlement patterns are such that HSR should be more viable, and as Marshall observes, "[t]here is no question that the Northeast has the need, demand and potential for true high-speed rail."

Sounding a skeptical note in the final article, "Money Train: In California, Obama's Grand Vision of America's High-Speed Rail Future Is Confronting Reality," Philip Klein takes aim at a planned HSR line connecting Los Angeles and San Francisco. Pointing out a host of impediments, from community opposition to exorbitant construction costs, Klein wonders if the project is worth the steep price.

Bullet Trains for America?[*]

By Mark Reutter
The Wilson Quarterly, Autumn 2009

Over the rice paddies and tin-roofed villages of coastal China, thousands of sleek, prestressed concrete bridge pylons are rising into place like a giant row of dominoes, at a pace that rivals the frenzied advance of the track layers who completed America's first transcontinental railroad in 1869. The "transcon," which linked the existing railroads of the East with California, was considered the greatest construction feat of its era. As more track was laid across the United States, wagon trains and stagecoaches faded from the scene, and railroads became the 19th-century's dominant carrier of passengers and goods.

China has embarked on a program that at first glance looks like a return to the past, but is viewed by government planners as vital to the country's fast-growing economy. After two decades of highway construction, the focus has shifted to public transportation, with the equivalent of $1 trillion allocated to expand and improve the railway network. The most audacious element of China's plan is to build 8,000 miles of high-speed railways by 2020. The first segment is already under construction between Beijing and Shanghai: 820 miles, comparable to the distance from New York to Chicago. When the line opens in 2012, trains on elevated rights of way will race at speeds as fast as 235 mph between the two cities, cutting the trip time from 12 hours to four and a half. Eventually, China wants to connect its rail network to a "supertrain" line to Europe that, carrying both passengers and export goods, would help secure the nation's future as a global powerhouse.

Since entering service in Japan in 1964, fast trains have been gaining in speed and popularity. The first Shinkansen, or bullet train, traveled between Tokyo and Osaka at a maximum of 130 mph. The latest-generation Shinkansen runs at a top speed of 188 mph, and its ancestor is now in a museum. France launched Europe's first high-speed railroad between Paris and Lyon in 1981, the Train à Grande Vitesse ("train of great speed"), better known as the TGV. Today, trains doing 125 mph or more zip across 13 European countries as well as Russia, South Korea, Taiwan, and Turkey. In the Middle East, Saudi Arabia recently let contracts for a European-style supertrain between the western port of Jeddah and the religious centers of Mecca and Medina, while Israel has a new Tel Aviv-to-Jerusalem line in the works and Iran is upgrading its main lines out of Tehran to standards exceeding 120 mph.

By these measures, America's passenger trains are slowpokes. Even Amtrak's self-declared high-speed Northeast Corridor between Boston and Washington does not qualify as high speed by world standards (defined by the International Union of Railways as regular operation at or above 155 mph on new or renovated track or 124 mph on older track). The Acela Express reaches 150 mph on a short stretch of reconditioned track in Rhode Island, but otherwise is forced to go much slower because of aging infrastructure. Overall, Acela trains average only 67 mph between Boston and New York City. South of New York, Acela trains operate at atop speed of 125 mph and an average of 77 mph. Compare this with the 217 mph maximum and 146 mph average of Spain's 386-mile line between Madrid and Barcelona, and the gap between U.S. and European railroads becomes apparent.

This gap only widens in the rest of Amtrak's 22,000-mile nationwide network. Outside of the Northeast Corridor, there are only four routes where Amtrak trains can run faster than 79 mph: Los Angeles-San Diego, New York City-Albany, Philadelphia-Harrisburg, and a 100-mile segment in Michigan. Elsewhere, trains are restricted to 79 mph because locomotives and track are not equipped with signal systems that prevent collisions. After accounting for speed-restricted curves, snail-like crawls through junctions, stops for opposing trains, and other obstacles thrown in their path, Amtrak trains average no better than 50 mph between terminals—and much less if unscheduled delays are counted. The result is that train service is slower today than it was in the 1940s, when "streamliners" touted for their speed—such as the Super Chief, 20th Century Limited, Denver Zephyr, and Hiawatha—routinely topped 90 to 100 mph between station stops. While the rest of the world has advanced, America's passenger rail has stalled, if not reversed direction.

If President Barack Obama has his way, American passenger rail will pick up speed again. Earlier this year, he called for the creation of a national high-speed rail network. The idea is not to lay track coast to coast, but to focus on heavily populated corridors where short distances between cities let fast trains compete effectively with cars and planes. President Obama allocated $8 billion from the economic stimulus package and requested $5 billion more from Congress through 2014, which would be used as seed money for improved rail service.

Ten corridors, ranging in length from 200 to 600 miles, have been designated as potential high-speed routes. These routes would serve city clusters that currently have no through passenger service (such as Miami-Orlando-Tampa), as well as corridors that have built up ridership on conventional Amtrak trains (such as New York-Albany-Buffalo and San Diego-Los Angeles-San Luis Obispo). State governments have subsidized Amtrak trains for years, but attempts to add faster and more frequent service have been thwarted by lack of money and, sometimes, by resistance from Amtrak itself. As an Illinois state senator in the 1990s, Obama witnessed firsthand the state's frustrating attempts to refurbish Amtrak's slow and threadbare service between Chicago and St. Louis.

The $8 billion program bypasses Amtrak and will provide funds directly to selected state and regional agencies. The plan follows the precedent of the Interstate Highway System, initiated by President Dwight D. Eisenhower in the 1950s, for which states planned and built the highways according to standards set by the federal government, which picked up 90 percent of the tab. For high-speed rail, though, private investors will also be sought. "This is not some fanciful, pie-in-the-sky vision of the future," Obama observed in introducing his plan. "It's been happening for decades. The problem is it has been happening elsewhere, not here."

While maybe not a pie-in-the-sky project, instituting high-speed rail—or even getting train speeds back to 1940s standards—will be a tall order requiring years of commitment and vastly more than $13 billion to pull off. Given Americans' well-known penchant to jump in a car or head for the airport to get where they're going, how realistic is Obama's plan?

Backers cite many gains to be reaped: relieving traffic congestion, promoting economic development, improving safety, and creating jobs for the tens of thousands of workers who will construct and operate the system. Saving energy and cutting greenhouse-gas emissions also are key selling points. Diesel-powered trains use 27 percent less energy per passenger mile than cars and 21 percent less than airliners, according to the Oak Ridge National Laboratory. If high-speed rail lines were operated with nonpolluting electric locomotives, they could reduce carbon dioxide emissions by as much as two million tons annually, according to the Center for Clean Air Policy. "Not since the implementation of the Interstate Highway System have we been afforded such a momentous opportunity to change how this country moves forward," Edward G. Rendell, governor of Pennsylvania and chairman of the National Governors Conference, told Congress last June.

Not so fast, say critics. According to Randal O'Toole of the Cato Institute, a Washington think tank, high-speed rail is a "mirage" that would do little to reduce highway traffic congestion or improve the environment. He and others argue that most congestion involves travel within cities rather than between them, and O'Toole contends that high-speed rail in any event "won't take more than three or four percent of cars off the highways it parallels." At best, supertrains would replace commuter airlines, and at worst, the lines would cause a long-term drain on public finances at a time when the United States is in dire fiscal straits.

"Taxpayers and politicians should be wary of any transportation projects that cannot be paid for out of user fees," O'Toole warns. But roads and airports are paid for only in part by those who use them through gasoline taxes and other user levies. For example, airline ticket tax receipts cover airport construction costs, but the costs of safety measures and a portion of air traffic control—more than $2.5 billion per year—are subsidized by all taxpayers, including those who never fly. Intercity trains are likewise subsidized out of tax revenues from the whole population. Last year, Amtrak earned 72 percent of its costs from ticket receipts and received about $1.2 billion in direct federal subsidies.

While there are important issues to be debated about the costs and benefits of high-speed railroads, the numbers on both sides of the equation are notoriously slippery. And longer-term benefits can be impossible to anticipate. Yet the fundamental choice facing the United States is about the longer-term future. The decisions made today about transportation will literally shape the American landscape and economy for decades to come, and in ways that are difficult to predict with any precision. Even the most farsighted planners behind the Interstate Highway System did not anticipate the extent to which the new roads they built would help spawn a burgeoning suburbia, with its far-flung office and industrial parks, edge cities, and immense shopping malls and commercial areas.

High-speed rail doesn't simply proceed from point A to point B; it has the potential to energize the cities and towns where it stops in between. The normal practice is to locate intermediate stations in populated areas roughly 50 miles apart. In Europe, high-speed railroads have generated the most growth in provincial cities, as once remote districts benefit from their newfound closeness to hubs such as Paris and Berlin. In a century that will demand more compact, energy-efficient development, high-speed rail has the potential to establish a new superstructure for growth.

Specialists generally agree that high-speed railways earn a high percentage of their costs when carefully planned. But do fast trains repay their total investment? In a 2009 survey of high-speed lines in Japan and Spain, the U.S. Government Accountability Office reported that Japan's core Shinkansen routes fully repaid the initial investment and debt related to their construction. However, three high-speed lines built in the 1990s, when Tokyo was trying to stimulate the economy through very liberal funding of public works, recovered only 10, 52, and 63 percent of their construction costs through ticket sales. In Spain, the original high-speed line between Madrid and Seville has been profitable on an operating basis, but has not yet repaid its original construction costs. Other studies indicate that France's TGV system operates at a profit, while Germany's and Italy's high-speed trains share in the subsidies given to the railway system by the government.

Critics have rightly pointed out that attempting to extrapolate the economic performance of the Shinkansen or TGV in the United States is a dubious exercise. For one thing, expensive gasoline and expressway tolls make driving a much less attractive option overseas. On the other hand, critics have steadfastly refused to look at high-speed rail outside the context of Amtrak or to recognize the phenomenon of "induced" traffic—the traffic that seems to spring up, seemingly from nowhere, when superior technology is introduced. This capacity for growth is a recurring theme of modern transportation dating back to when steamships replaced sailing vessels and jet aircraft overtook propeller planes. The deciding factor is speed or fluidity of movement. And the more customers, the lower the cost per passenger mile.

The debate is confusing in part because few Americans know what fast trains really are, much less how they can best operate. Many people simply assume that Amtrak schedules would magically accelerate if, say, some bullet-shaped trains were placed on the track and their motors revved up.

It's not that simple. High-speed trainsets are only as good as their supporting infrastructure: right of way, track quality, and propulsion. Requirements for each component become more exacting as the speed of a train increases. Take the matter of track quality. Freight and conventional passenger trains can operate safely with relatively large discrepancies between the level of one rail and the other. The Federal Railroad Administration (FRA) permits a maximum discrepancy of 1.25 inches for 79 mph operation. But the French require discrepancies of no more than 0.16 inches for the TGV, and the FRA standard for 120 mph along the Northeast Corridor is 0.5 inches. "There is nothing impossible about such requirements," Louis S. Thompson, a former FRA official, has written. "Satisfying them is, however, expensive."

The freight railroads that own the track that Amtrak uses for its passenger trains (except along the Northeast Corridor, which Amtrak owns) have little incentive to upgrade track to high-speed standards. The current track works just fine for them. In fact, many railroads reconfigured their track after Amtrak was formed in 1971, relieving private railroads rocked by the bankruptcy of the Penn Central Railroad of the need to operate passenger trains. Banked curves, which kept the centrifugal force on passengers to a tolerable level at high speed, were flattened to better accommodate slow-moving freights. Amtrak now must brake for curves that streamliners once navigated with ease.

Because a railroad built today will probably still be in operation 150 years from now, rights of way should be engineered for maximum speeds from end to end. Overseas, where many high-speed trains run on dedicated tracks, they are. (In Japan and most of Europe, passenger trains are dominant and railroads carry a relatively small proportion of all freight—the reverse of the situation in the United States. Europe's freight railroads are hampered by national differences in signaling systems and other technologies.) The sharpest curves permitted for trains operating over 170 mph are a close approximation of a straight line. Grades are typically restricted to one percent, or a one-foot rise or fall per 100 feet of distance. "Together, the limits on curvature and gradient mean that high-speed rail requires extensive land acquisition and expensive cutting, filling, bridging, and tunneling, especially in hilly areas," Thompson noted.

Expanding such corridors through heavily populated areas presents environmental hazards and NIMBY (not in my backyard) challenges, not to mention costs ranging up to $50 million a mile.

In deciding which high-speed projects to fund, the federal government will need to choose between two types of propulsion systems. Freight railroads—and, by extension, Amtrak—use diesel locomotives. The initial cost of diesel power is far less than that of electric propulsion, which requires overhead electric lines and trackside transformers. While today's diesel locomotive produces 70 percent less pollution than its 1980 counterpart, electric power uses less energy, emits no pollution, and offers faster acceleration than diesel engines, an advantage at high speeds. Electricity is the standard propulsion for the TGV and other overseas railways, but in the United States it is only available along the Northeast Corridor. Electrifying just one of the high-speed corridors proposed for the Midwest, the 300-mile Chicago-St. Louis line, would cost $1.2 billion, according to TranSystems, a transportation planning firm.

So how much is speed worth? Without a doubt, fast trains attract more passengers. A general rule of thumb is that every minute saved in transit is likely to generate one percent more customers. But does there come a point where increments of speed are not worth the extra outlays of money? This question will be critical to the ultimate success of President Obama's railroad plan.

The administration has outlined a "three-track investment strategy" to divide up the $8 billion in seed money. The first track is not really high speed at all. It would provide money for incremental "shovel-ready" projects that could nudge up the speed and frequency of conventional diesel-powered trains. Many states are angling for these federal dollars. The goal of North Carolina's grant submission is 85 mph service between Charlotte and Raleigh as part of the effort to implement faster rail service along a 450-mile corridor between Washington, D.C., and Charlotte. Oregon wants to upgrade track and crossings so trains between Portland and Eugene can average 65 mph.

The second type of project, known as "emerging high-speed rail," would boost train speeds to the 110–125 mph range on existing freight lines. The Association of American Railroads currently requires dedicated track for passenger trains running at 90 mph and over. An agreement relaxing this policy to permit shared-use trackage could reduce expenses. Still, retrofitting freight lines will not be easy or cheap. A coalition of mid western governors hopes to use stimulus money to develop lines out of Chicago with train speeds of 110 mph. Wisconsin plans to rehabilitate a rail link between Milwaukee and Madison at a cost of $600 million. This is part of a plan to reduce a trip between Chicago and St. Paul, now eight hours, to five and a half hours. Refurbishing the Chicago-St. Paul route would cost $2 billion, not counting the price of new trains.

What has stirred the most excitement and controversy is the development of trains capable of 200 mph over exclusive, built-from-scratch lines. The most ambitious project comes from the state that gave rise to the freeway. Trains have made a steady and little-noticed comeback in the Golden State. The San Diego-Los Angeles-San Luis Obispo corridor is the nation's second busiest intercity rail line, surpassed only by the Northeast Corridor. Last year, it carried three million riders.

Now the California High-Speed Rail Authority has developed plans for an 800-mile line between Sacramento and San Diego. Trains would operate at a top speed of 220 mph, making the trip between Los Angeles and San Francisco in 2 hours, 40 minutes; the line would attract as many as 100 million riders a year. Last November, California voters approved the sale of $9 billion in bonds for construction. But another $35 billion will be needed. With the state government mired in a fiscal crisis, California's ability to finance the project has been cast into doubt. The authority is seeking $1.3 billion in federal stimulus money to match hoped-for state aid in order to fund preliminary construction.

Several other states are vying for federal dollars for fast trains. In Florida, advocates are trying to revive plans for a TGV-type railroad linking Tampa, Orlando, and Ft. Lauderdale/Miami. Florida asked for $1.5 billion in stimulus funds to build the first leg between Tampa and Orlando International Airport, to be matched by $1 billion in private investment. In Texas, a fast train has been proposed linking Houston, San Antonio, and Dallas.

In 1955, as plans for the 40,000-mile Interstate Highway System were taking shape at the Eisenhower White House, *Fortune* magazine pointed out that "the administration has a highway plan with but one major flaw—it costs money." A huge amount of money, in fact. First estimated at $27 billion, the price of the interstate system soon ballooned to $40 billion (about $280 billion in today's dollars).

Attempts to pay for highways with tolls were successful only in the heavily traveled urban Northeast, where roads such as the New Jersey Turnpike had been completed. In the Midwest, a tollway between Pittsburgh and Chicago was financially viable, but in Texas, promoters of the Sam Houston Turnpike Corporation found it impossible to float bonds. About four-fifths of President Eisenhower's proposed interstate network was stopped in its tracks due to insufficient funds. Eventually, the administration and Congress developed a "pay as you go" system that relied on federal and state user fees on gasoline and other motor fuels to finance the program, and interstate construction got under way.

President Obama is faced with a similar challenge. Given the fiscal plight of states and the growing federal deficit, government alone probably cannot finance 10 high-speed corridors that ultimately might cost a total of $200 billion or more. The creative use of private capital will be needed to proceed. In Florida, for example, railcar equipment makers have pledged to help finance the first phase of the Tampa-Miami corridor, and private operators are expected to bid for the right to operate the line. In the 19th century, the federal government gave land grants to private investors to jump-start railroad projects. Similar grants of real estate or other benefits, such as access to rights of way along rail lines for building fiberoptic and utility lines, could help spur investment in high-speed projects.

In his effort to redirect America's transportation priorities, Obama said he was inspired by an earlier project that changed the course of the country. The transcontinental railroad was an example of "bold action and big ideas" during a period of "economic upheaval and transformation," he reminded a joint session of Congress in February. Despite the crushing costs of the Civil War, President Abraham Lincoln authorized a 1,700-mile railroad between Omaha and Sacramento. Because no state governments existed in the lands to be traversed, the federal government subsidized private businessmen, led by Leland Stanford of California and the Ames brothers of Massachusetts, by giving them public land as well as cash bonds for every mile of track completed.

Lincoln did not live to see the ceremonial golden spike driven into a crosstie in a barren corner of the Utah Territory on May 10, 1869, which joined the Union Pacific and Central Pacific railroads and ushered in what economic historian Walt Rostow called the "takeoff period" of the American economy. But the moment was captured by a telegraph operator who sent a message to a waiting nation that might be repeated someday if Obama's railway initiative gains traction. It simply said, "DONE."

MARK REUTTER, *a former Woodrow Wilson Center fellow, wrote "The Lost Promise of the American Railroad," which appeared in the Winter 1994 issue of the* WQ, *and can be found online at www.wilsoncenter.org/Train.pdf. He edited* Railroad History *for eight years and is the author of* Making Steel: Sparrows Point and the Rise and Ruin of American Industrial Might *(2005, rev. ed.).*

All Aboard?*

Bullet Trains May Be Sidetracked, But Not Commuter Rail

By Peter A. Harkness
Governing, January 2011

Why does this country have such a problem with trains that carry people?

It isn't just the intercity rail system, which is something of a joke everywhere but the Northeast Corridor—and even there it's a poor performer by international standards. But it's also the aging "heavy" rail mass transit systems in cities like Atlanta, Chicago and Washington, D.C., which are facing huge bills for deferred maintenance, forcing service cutbacks and fare increases—even as ridership has risen to levels not seen in more than 50 years. Light commuter rail has caught on in numerous cities, but there are problems in securing rights of way and controlling costs. Oddly, the bright spot is freight. The nation's railways don't handle passengers well, but they do a bang-up job hauling stuff—much better than most other countries. Most passenger trains must travel on lines owned by thriving railroads, contributing to the congestion many commuter rail systems are experiencing.

The Obama administration is more sympathetic to rail transit than its predecessors. It proposed a historic expansion of the rail passenger system, including building a national high-speed network of bullet trains with an initial $8 billion down payment in stimulus money (with more promised) to a few states for some modest projects to get things going.

The problem is that the newly elected Republican governors of states where much of the money was supposed to go—like Ohio and Wisconsin, and maybe Florida—don't want it, at least not for high-speed rail. They'll gladly take it for auto infrastructure like roads, bridges and highways. But U.S. Transportation Secretary Ray LaHood, a former Republican congressman from Peoria, Ill., won't agree to that: It's accept rail or hit the trail, and the money will go to states that want it.

Recently the greater New York area was stunned by New Jersey Gov. Chris Christie's decision to pull his state out of a long-planned project—described as the largest public transit program in the country—to build a second rail tunnel beneath the Hudson River to ease the commute by 45 minutes for Jersey residents who work in New York City. With substantial overruns, it was estimated to cost as much as $13 billion. Christie's state was on the hook for $2.7 billion, plus the added costs for its share of the project, which already is under construction. Much is at stake, including 6,000 construction jobs.

Making significant improvements in rail service in this country seems like a no-brainer. Ridership is increasing. The highways and airways are overburdened. It's far more energy efficient and cleaner, and compared to cars, it's safer. If done right, it can be one of the most effective economic development tools available. But it's also very expensive and requires a sustained commitment over many decades. And right now, governments are deep in debt.

Critics of Obama's high-speed rail plan make several points. The project will cost far too much in initial outlays and subsidies to justify the benefits, siphoning off the funding of worthier programs, including commuter mass transit. The United States has become a suburbanized society, sprawling over a large land mass, with only a few places having sufficient population density to warrant intercity rail service. To be successful in any area except the Northeast Corridor, high-speed trains would have to make too many stops, and therefore would be too slow to compete.

Given the political changes in the new Congress and in many states, it's hard to imagine that we'll see many bullet trains whizzing through our future. But that doesn't necessarily mean that all is lost for rail advocates. The incoming chairman of the U.S. House Transportation and Infrastructure Committee, Florida Republican John Mica, is outspoken in his opposition to the administration's plan, which he claims is likely to lead to many "slow-speed trains to nowhere." But he does support what he calls "a better directed high-speed rail program."

What's that likely to mean? The first demonstration grant of $1.25 billion linking Tampa to Orlando may go through, despite the concerns of high-speed rail advocates. They believe that it could set back their cause because there are five stops along its 84-mile route, so it will only cut 30 minutes off the trip by car.

Mica and others argue that the best way to spend limited resources now would be to first invest in improving dense corridors—like in the Northeast, where Amtrak's Acela service already exceeds the airlines in passenger volume between Washington, D.C., and New York City, even though track conditions prevent the trains from maintaining high speeds. The Los Angeles-to-San Francisco corridor also might qualify.

Beyond that, most advocates and critics can agree on the need to make strategic improvements where there now are serious bottlenecks in the system that delay both freight and commuter trains, like around Chicago or New York.

Meanwhile, the political jockeying over passenger rail money continues. California and other states will receive more federal funds because Ohio and Wisconsin are turning them down, prompting the *Los Angeles Times* to chortle, "This is sort of like turning down a free car because you don't want to have to pay for gasoline and insurance."

Leaping into Light Rail[*]

Cities in the South and West Are Writing a New Chapter in the History of Transit

By Zach Patton
Governing, July 2008

"Hey! What do you know? It's cold!"

Rick Simonetta isn't talking about the temperature in downtown Phoenix, which today is 92 degrees in the shade. What's cold is the water he's sipping from a fountain on the platform of an outdoor train station. Right now, the platform is eerily devoid of passengers, its modern, sage-colored ironwork glinting in the Southwestern sun. But come December, when Phoenix opens its first light-rail line, this station will anchor a huge regional transit system that will stretch north to Glendale and east to Mesa and Tempe. It's a $1.4 billion, 20-mile catapult into transit—no other light-rail system in the country has been so large right from its inception.

As Simonetta, the system's CEO, inspects the gleaming new stations, he's focused on the details. Phoenix, he notes excitedly, will boast—make that boasts already—the first light-rail stations anywhere with chilled water fountains. Simonetta figures that will be a necessity if he's going to persuade drivers to get out of their air-conditioned cars and stand in the heat waiting for trains. To that end, other touches at the stations include sweeping sail-shaped sun shades, tilted in a way that keeps at least 40 percent of the platform shaded at all times, and sand-colored concrete that deflects heat rather than absorbs it. Simonetta can easily envision this station teeming with commuters headed to work, college students on their way to class and other residents coming downtown to shop, catch a concert or watch basketball's Suns or baseball's Diamondbacks play.

Owing to the heat, light-rail stations in the Valley of the Sun may look and feel a little different than those in other cities. Still, the story of another light-rail grand opening has a familiar ring to it. That's because cities around the country have been investing in new systems like never before. That's especially true in the land of Sun Belt sprawl, where the phrase "public transportation" has historically been something of an oxymoron. Austin and Norfolk will launch their own light-rail lines in the next couple of years, joining recently opened systems in places such as Charlotte, Denver, Houston and Salt Lake City. Add Phoenix to the list and you could argue that this pack of cities forms something of a graduating class. They're all moving beyond a car-bound past for a new future with light rail—and they are all making the change more or less simultaneously.

The current spate of light-rail openings represents the culmination of a decade of both new and recycled ideas about urban planning, transportation choices and how to finance big infrastructure projects. Light rail is enjoying something of a national moment, akin to the subway boomlet 40 years ago that brought MARTA to Atlanta, BART to the Bay Area and Metro to Washington, D.C., and its suburbs. There are some key differences between that era and this one, though. For one thing, the light-rail systems being built today are expressly intended to catalyze transit-oriented development in ways that the last generation of rail investment hadn't envisioned. Plus, Americans seem more ready to ride the rails now than they were in the 1960s and '70s, when their love affair with driving and suburban-style living was still fresh.

That's thanks in no small part to the price of oil. As gas prices soar over $4 per gallon and commuters struggle with ever increasing traffic congestion, drivers are turning to public transit in record numbers. The American Public Transportation Association reports that Americans took 2.6 billion trips on public transportation in the first three months of 2008—nearly 85 million more trips than last year for the same time period. In total, 10.3 billion trips were taken on public transportation last year—the highest number in more than 50 years. By far the biggest increase was in the number of trips on light rail, which saw more than a 10 percent jump in ridership. Several cities' light-rail and streetcar systems grew even faster. Baltimore. Minneapolis, St. Louis and San Francisco all experienced significant growth in passenger loads.

BACK TO THE FUTURE

The current national interest in light-rail transit is, of course, something of a full-circle return to the approach from more than a century ago. Before the rise of the automobile, most cities of any size had streetcar systems. According to the Center for Neighborhood Technology in Chicago, the number of city streetcar systems rose from just one in the whole country in 1885, to at least one in every city of more than 5,000 people by 1902. Scott Bernstein, the center's president, says, "It was the fastest growth in mass transit in our nation's history."

At that time, these systems were privately run and highly profitable. Many commuters worked six days a week, returning home for lunch in the middle of the day, meaning the trains were full throughout much of the week. Times changed.

As work patterns shifted, transit trips became less frequent, and as more families later acquired automobiles, ridership dwindled even further. Cities took over the struggling systems, operating them at a loss. By the 1950s, most places had pulled up their streetcar rails and converted to buses, which required a tiny fraction of the upfront capital cost of rails. The era of the auto was in full swing.

In the midst of this era, though, some cities embarked on large-scale mass transit projects. In the '60s and '70s, Atlanta, Baltimore, Miami, Washington, D.C., and the San Francisco Bay Area all built heavy-rail lines. The systems were commuter workhorses, designed to truck people in from the suburbs to the central city. It was a boom time for rail, but as those cities built out their systems, the nation's capacity for subways and heavy rail seemed tapped out.

Today's focus is light-rail transit—trains that typically run at street level, with the traffic. Light-rail platforms are more akin to bus stops than subway stations. For these reasons, light rail costs a fraction of what heavy rail does. The city of San Diego launched the first modern light-rail line in 1981. But it was only after Portland, Oregon, demonstrated how light rail could drive development patterns and Dallas showed that trains in the Sun Belt could attract solid ridership that more cities began passing local-option sales taxes to pay for systems of their own. Light rail became the new transit zeitgeist. "Over the last 10 years, but really since 2000, light rail has exploded, both in terms of cities' interest in building it and in terms of ridership," says Jason Jordan, director of the Center for Transportation Excellence, a nonpartisan research center on transit policy. A lot of these cities are growing very fast, and there are definitely concerns about congestion and pollution. But there's also something a little bit more ephemeral to this. It's about creating what a world-class city looks like."

Yet light rail has always had its critics, especially among fiscal conservatives and libertarians who say it simply costs too much to move too few people. One persistent critic is Randal O'Toole, a senior fellow at the Cato Institute who writes a blog called *The Antiplanner*. Responding to the news about rising transit ridership, O'Toole wrote, "Let's say transit continues to grow by 2.1 percent per year and driving grows only at the rate of population growth, or 1.0 percent per year. Then transit will grow to 10 percent of total urban motorized travel after a mere 178 years. I can hardly wait. Until then, transit is pretty irrelevant in any cities not named New York."

EVOLVING VIEWS

Libertarian objections in the transportation arena have deep roots in the Phoenix area. Right up through the 1970s and '80s, if you were driving through Arizona on Interstate 10, the road would have run out when you got to Phoenix. At the very same time that some other cities were busy building expensive subways, the Phoenix area so opposed government intervention that even an Interstate highway was considered suspect. "It was that whole mindset that we think, 'Oh, that's not a role government should have,'" says Rick Simonetta. "We were late in the game of Interstate building, late in the game of transit, and late in the game of realizing there is a governmental role in building infrastructure."

The "Phoenix gap" may be an extreme case. But it's a reminder of how much some cities in the South and West have changed their worldview in order to support big investments in light rail, says Jordan. "You look at cities like Phoenix, Charlotte, Dallas. These are not just cities without a history of transit. These are cities that were considered downright hostile to transit."

That hostility hasn't completely disappeared. As Charlotte prepared to open its first light-rail line last fall, public-transit opponents mounted an aggressive ballot campaign to cut funding for the project. Although voters in November ultimately opted to continue funding, the anti-transit campaign gave city officials several months of nervous nail biting.

Or, take Norfolk, where a new light-rail system known as "The Tide" is set to open in 2010. The route there begins at a medical center on the west side of town and runs 7.4 miles east through Norfolk—until it dead ends at the border of the city of Virginia Beach. Although the line was originally meant to be a regional system, Virginia Beach voters opted out of the plan in 1999 and never looked back.

The Phoenix area, too, has seen its share of pushback. The light-rail project was officially organized in 2002, but that capped more than a decade of false starts, as citizens repeatedly voted down tax increases to fund it. Although public opinion in the region now seems to have largely turned, there remain pockets of deep opposition. In the tony suburb of Scottsdale, there was an outcry against even considering whether light rail should come through town. The city eventually decided to join the regional board overseeing light rail, at an annual cost of $50,000, but that still rankles Councilman Bob Littlefield. "The whole push for light rail isn't about transportation," Littlefield says. "It's about development and urbanization— development that would be out of character for Scottsdale."

For this class of regions that are just coming around to the idea of public transit, opposition isn't the only problem. There's also the challenge of building and operating a rail system in an area where drivers and pedestrians simply aren't accustomed to sharing the road with a train. That was a huge problem in Houston when that city opened its first light-rail line several years ago. The first train-car collision occurred during the testing phase, before the system was officially open. By mid-2004—six months into operations—Houston's METRORail set a new record for most accidents in a year. Critics dubbed the Houston lines the "Streetcar Named Disaster" or the "Wham-Bam-Tram."

Since then, Houston and other cities have put more of an emphasis on educating drivers and pedestrians. Phoenix actually overhauled its blueprints in response to the accidents in Houston. "We were 65 percent into our design process and we went back to the drawing board," says Maria Hyatt, assistant to the Phoenix city manager. The city redesigned many of the intersections where trains and automobile traffic would meet. New features were added, such as large sidewalk planters to discourage pedestrians from walking into the path of a train. "All of the changes were a result of going to Houston and seeing where their accidents were occurring and why they were occurring."

CONTEXT IS KING

As Rick Simonetta continues walking around downtown Phoenix, pointing out the features of the light-rail stations, it's a little hard to hear him over the din from all the construction sites. Downtown Phoenix is in the middle of a $6 billion building boom—fueled in part, Simonetta says, by the coming of light rail. Mixed-use residential developments are shooting up everywhere, including a $1 billion project on the site of a former parking lot The city has just tripled the size of its downtown convention center. Two blocks away, there's a large new Sheraton hotel, built with city funds.

Next to the hotel is a new downtown campus for Arizona State University. Three years ago, the campus didn't exist. This fall, ASU expects to have 8,000 students based downtown—and the university plans to eventually boost that enrollment to 15,000. The school is building classroom facilities, dorms and a library, and the hope is that students and faculty will use the train to shuttle between the downtown campus and the main campus, which has a stop on the line in Tempe. "ASU is the first success story of our light rail," says Phoenix deputy city manager Tom Callow, "and the system hasn't even opened yet."

If there's any single thing that separates the current light-rail boom from the subway-building era of the 1970s, it's this notion of context. It's the idea that transit systems aren't just for moving commuters from their homes in the suburbs to their jobs downtown. Transit today is seen as a tool to drive development and to help shape the way a city will grow. "There is the understanding now that transit can't succeed in a vacuum," says Jason Jordan. "The cities that are successful are the ones that really work to integrate transit with the city and to encourage transit-oriented development." Contrast that notion, Jordan says, with one of the mid-century systems, such as Atlanta's MARTA, which was designed with a total emphasis on moving large numbers of people into downtown. The notion that transit should interact with the urban environment was an afterthought, although Atlanta now is working to boost transit-oriented development. "It was this idea that you could just plop a system down into a non-transit-supportive environment and it would work just fine. But you're going to have problems. The current investment in light rail, then, isn't being undertaken in isolation. The light-rail boom is part of a much broader movement toward creating community."

In the current light-rail moment, cities finally seem to be realizing the full potential for transit to catalyze urbanism. But the current moment could be over soon. After this graduating class of systems finishes coming on line, it will be sometime before they're joined by any new cities. For one thing, the skyrocketing costs of construction materials are making capital investment in light rail much more expensive than it was only a few years ago. And in a weak economy plagued by a housing crisis and a credit crunch, cities may be skittish about asking voters to fund multibillion-dollar rail projects.

The biggest reason why this transit chapter may be coming to a close, however, is because the federal government isn't as keen on light rail as cities are. All federal money for light rail is approved through the Federal Transit Administration's New Starts program. Unlike road projects, light-rail systems seeking federal funds must compete against one another—and against other forms of transit that federal criteria have tended to favor, such as bus rapid transit. New Starts doesn't take into account development potential, creating a sense of place, or reducing harm to the environment. Rather, funding is based on the ability of a system to replace car trips with transit trips for the least amount of money.

That may well change in 2009, depending upon the priorities of the next president. For now, though, cities must rely very heavily on their own funds, and on asking voters for more money. It's a situation that Phoenix's Tom Callow says may threaten light rail's potential. "I do think there's a national moment, but I'm not sure the nation's ready to seize it. I think we should, but it's going to be hard." On a bookshelf in Callow's office, high above downtown Phoenix, there's a collection of miniature cars—Callow was the city's director of street transportation for years before he began working on the light-rail project. "I've been a car guy all my life," he says. "But now I've seen the light."

Accelerating Amtrak[*]

Will Superfast Train Service in the Northeast Ever Happen?

By Alex Marshall
Governing, August 2010

Many great ideas citizens thought would never happen actually did occur—the Berlin Wall fell; welfare as we knew it ended; and Times Square was revitalized.

The latest great idea that people say will never happen—but it just might—is building true high-speed rail in the Northeast, from Boston down to New York, through Philadelphia, Baltimore and Washington, D.C. This region has roughly 55 million people in densely populated cities that are 50 to a few hundred miles apart, and it has a tradition of train travel with mass transit use: Amtrak trains, pricey and unreliable, are still packed.

Yet somehow—as with Times Square when it was still a cesspool of porn and crime, or East Germany when it was under the Soviet Union's thumb—people in the Northeast shrug their shoulders and tend to say that nothing's going to change. They've lived so long with average quality intercity train service that they don't ask for or expect the best.

Meanwhile, countries like France, Japan, Germany, Spain, China, South Korea, Brazil and Argentina—even Morocco and Vietnam—either are proceeding with new high-speed lines or enjoying extensive networks that are already built.

But in this age of infrastructure, plans are afoot, even in the Northeast: This year, 11 states, Washington, D.C, and Amtrak—with New Jersey's NJ Transit acting as the formal applicant—asked the Federal Railroad Administration for $18.8 million to conduct a high-speed rail study that encompasses both the immediate repair needs and an examination of longer-term needs through 2050.

The proposal, called the *Northeast Corridor (NEC) Multi-Modal High Speed Rail Improvement Plan*, would include a Programmatic Environmental Impact Statement, an essential legal first step for any large interstate transportation project.

This effort is expected to also involve the commuter and freight rail operators using the Northeast Corridor and its tributary rail lines.

Two questions above all hover around these studies: What is truly high-speed rail? And how much should be spent on it? Bureaucracies, like people, must judge how high to aim. Is it realistic to try for true high-speed rail?

Or is simply higher-speed rail enough? Already Amtrak's Acela Express trains hit 150 mph for a short section between Boston and Providence, R.I., but average travel times are less than 100 mph for the whole route. To get true high-speed rail with trains traveling at 180 mph and averaging more than 150 mph, you need separate right of way and new tracks—a difficult thing in the crowded Northeast, not to mention expensive.

However, there is a plan to do just that, and it's worth looking at to get a sense of what's possible. A largely separate, new-track high-speed rail system could be built between Washington, D.C., and Boston for a price tag of $98 billion, according to a University of Pennsylvania graduate planning studio led by Robert Yaro of the Regional Plan Association (where I am a senior fellow) and Marilyn Jordan Taylor, dean of the University of Pennsylvania School of Design.

The plan is appealing in both its ambition and apparent feasibility. North of New York City, for example, rather than attempting to carve out new tracks through the chaotic suburbs, the high-speed lines would travel under Long Island Sound through a 20-mile, three-tube tunnel to New Haven, Conn. From there, it would continue on existing rail right of way to Hartford, Conn., and then turn northeast onto the median of Interstate 84 from Hartford to Boston, thus obtaining much of the needed corridor for new dedicated tracks in one fell swoop. South of New York, a combination of new tracks and tunnels built on utility and other accessible rights of way, as well as some existing tracks, would be employed.

This work would enable trains to travel from Washington, D.C., to New York in 90 minutes instead of the current top time of two hours and 45 minutes, and trains between New York to Boston would take less than two hours instead of the current three and a half hours. The new tracks would allow for greater capacity, giving the region a broad mix of local and express trains. Smaller cities would for the first time gain from high-speed travel, while the huge boost in capacity would free up track space for commuter railroads in Baltimore, Philadelphia, Boston and New York, thus expanding their potential.

True high-speed rail service, the plan estimates, would revitalize small and large cities, concentrating growth there and helping make possible an environmentally improved and more pleasant lifestyle. The economic impact can't be understated: Faster, more reliable movement of goods and people in such a densely populated area will allow the economy of this mega-region to grow and flourish for decades to come.

Of course, this is just a plan from a graduate school class. But both Yaro and Taylor are key players in regional and national planning, and top professionals from the United States and Great Britain were included on the project. Taylor and Yaro hope their study will influence the formal studies being conducted.

There is no question that the Northeast has the need, demand and potential for true high-speed rail, and given what's at stake, the price tags being thrown around—from $50 billion to $100 billion—are not that large.

Maybe it's time to do it, just like those other things people thought couldn't be done.

Money Train[*]

In California, Obama's Grand Vision of America's High-Speed Rail Future Is Confronting Reality

By Philip Klein
The American Spectator, March 2011

"Within 25 years, our goal is to give 80 percent of Americans access to high-speed rail," President Obama declared in his State of the Union address, making it the most ambitious element of his vision for "winning the future."

Invoking national pride, Obama mused that "America is the nation that built the transcontinental railroad, brought electricity to rural communities, constructed the interstate highway system." Sadly, he lamented, the U.S. now lags behind Europe, Russia, and China in modern transportation infrastructure.

If the nation met his goal for high-speed rail adoption, he said, "This could allow you to go places in half the time it takes to travel by car. For some trips, it will be faster than flying—without the pat-down. As we speak, routes in California and the Midwest are already underway."

To most Americans, the passing reference to California was likely an afterthought, lost amid all the dreamy rhetoric of rebuilding the nation. But upon closer inspection, the state's proposed high-speed rail system serves as a perfect example of the gap between the promise of transformational liberalism and the reality of big government. Taxpayers everywhere should pay attention, because the project has already been granted $3.2 billion in federal funds, mostly through Obama's economic stimulus package—and its backers hope to gobble up billions more over the next decade.

The $43 billion transportation project to link Los Angeles to San Francisco with a bullet train by 2020 would be considered grandiose during the plushest of times, yet it's being pursued during an era when governments at all levels are mired in deep fiscal crises. The plan has been subject to a series of scathing reports by independent analysts, raising concerns about everything from its cost estimates to its business model. The University of California at Berkeley has questioned its lofty ridership projections. And even the *Washington Post* has editorialized against it.

Although voters in the financially strapped Golden State approved a ballot measure in 2008 authorizing up to $9.9 billion in bonds to build the rail system, the project has encountered a lot of opposition as it has progressed. Several cities are suing to prevent the trains from tearing through their downtowns. Farmers are worried that the tracks will carve up their land. Some environmental groups normally predisposed to supporting high-speed rail have turned against the proposed route, fearing its effects on undeveloped areas. When the High-Speed Rail Authority announced that the initial section of the line would be built in the state's less inhabited Central Valley region, many were puzzled as to why they didn't begin by connecting large cities with more potential riders. As a result, critics dubbed it the "train to nowhere."

"The cost projections are overly optimistic," Wendell Cox, a public policy consultant and coauthor of a critical report for the libertarian Reason Foundation, says. "The ridership projections are absolutely crazy. The thing will have no impact on highway traffic and will have little or no impact on the amount of planes in the air. This project really defines the term 'boondoggle.'"

The project will rise or fall based on federal commitments, a reality that spurred California state senator Doug LaMalfa to visit Washington in early January to make a rather unusual request for a state legislator.

"I know they're not used to this, but I asked them to stop sending us money," LaMalfa, a Republican, said. "Please stop sending us money . . . When they send us money, it actually costs us money."

So, at a time of unprecedented debt, why are the state government and the Obama administration still committed to the high-speed rail project? Why are planners starting the construction in a tiny, almost-unknown town outside of Fresno rather than in a major population center? And is there any chance of putting the brakes on the project?

Bringing high-speed rail to America has been a decades-long dream for liberals, who have long envied Europe's extensive rail system. Building a high-speed rail network, they hope, would move the nation away from automobiles and reduce pollution. It has the added bonus of being a massive, centrally planned public works project. The problem is just because rail has worked elsewhere, that doesn't mean it makes sense here.

"We're not like Spain or France, where the population densities are a lot higher, and the cities are not as spread out," Ken Orski, a former transportation official in the Nixon and Ford administrations and publisher of the newsletter *Innovation Briefs*, says. "So you can connect cities like Barcelona and Madrid or Paris and Marseilles easily."

In addition, large European cities have "distribution systems," meaning that when passengers arrive at a station, they can get where they need to go by public transportation or walking, without a car. By contrast, in a city like say, Fresno, a person would be stranded without one.

"So people who are saying 'Look at Europe, why can't we be like Europe?' I don't think they really realize the difference between our geographic and demographic conditions and theirs," Orski says.

The only place where high-speed rail could theoretically make sense would be the Northeast corridor from Washington to Boston, which would pass through Baltimore, Philadelphia, and New York. The problem is, Orski explains, it's likely "50 years too late," because the area along that route is already densely populated and developed, making it cost prohibitive to acquire right of way.

Nonetheless, when candidate Obama began speaking glowingly about high-speed rail during the Democratic primaries, liberals only cheered him on.

"This is something that we should be talking about a lot more," Obama said in a May 2008 event in Beech Grove, Indiana. "We are going to be having a lot of conversations this summer about gas prices. And it is a perfect time to start talk about why we don't have better rail service. We are the only advanced country in the world that doesn't have high-speed rail."

A few months later, in Milwaukee, he reiterated that "the time is right now for us to start thinking about high-speed rail as an alternative to air transportation, connecting all these cities and think about what a great project that would be in terms of rebuilding America."

In February 2009, Obama made sure that the economic stimulus package included $8 billion in funding for high-speed rail projects (he originally requested $10 billion).

"I put it in there for the president," then chief of staff Rahm Emanuel said in an interview with *Politico*, explaining how high-speed rail funding was inserted during congressional negotiations. "The president wanted to have a signature issue in the bill, his commitment for the future."

Secretary of Transportation Ray LaHood told National Public Radio that "this is the transformational issue for this administration when it comes to transportation. I think President Obama would like to be known as the high-speed rail president, and I think he can be."

In April 2009, Obama, with vice president Joe Biden (who has long boasted of riding the train from Wilmington to Washington each day when he was a senator), outlined a vision for building a high-speed rail system. Obama employed the same sort of big government jingoism as in this year's State of the Union address.

"There's no reason why we can't do this," Obama said. "This is America. There's no reason why the future of travel should lie somewhere else beyond our borders. Building a new system of high-speed rail in America will be faster, cheaper, and easier than building more freeways or adding to an already overburdened aviation system—and everybody stands to benefit."

He continued, "Now, I know that this vision has its critics. There are those who say high-speed rail is a fantasy—but its success around the world says otherwise."

In January 2010, when it was time to award grants for high-speed rail projects through the stimulus, California was the big winner—having received $2.25 billion in federal funds.

Back in 1996, the state legislature created the California High-Speed Rail Authority to develop an inter-city high-speed rail service. After several delays, the issue was taken to voters in 2008, who approved a ballot measure allowing the state to issue $9.9 billion in bonds for the purposes of planning and building the system. The measure received the support of 53 percent of Californians. Under the terms of the proposal, however, the state could only issue the bonds and spend the money by matching funding from other sources. Thus, it was fortuitous for the Authority that the approval of the ballot measure coincided with Obama's election and expanded Democratic majorities in Congress full of legislators hoping to use the economic crisis as an excuse to dole out hundreds of billions of dollars to their pet projects.

The federal government's largesse came with a number of strings attached that affected both the timeline and location of the initial section of construction on the high-speed rail line. In late October 2010, just a week before Election Day, the Obama administration raised eyebrows when it directed $715 million in additional money into the project, this time as a grant from the Federal Railroad Administration. Interestingly, the money was specifically directed toward building the first segment in the Central Valley, which would seem like a bizarre choice given its lower population density. Influential *Sacramento Bee* columnist Dan Walters pounced: "You'd have to be terminally naive not to believe that the splashy announcement, made personally by an Obama administration official in Fresno, was to help an embattled local congressman, Democrat Jim Costa, stave off a very stiff Republican challenge." Costa ended up being reelected by a margin of 3,000 votes, and wasn't declared the winner by the Associated Press until three weeks after the election.

Things took a more perplexing turn last December, when California's High-Speed Rail Authority announced its choice for the initial 65-mile segment of the line, connecting Borden and Corcoran. If you haven't heard of them, don't feel bad—neither have most Californians. Together the towns have an estimated population of 25,000. But as the *San Jose Mercury News* noted, Borden "is an unincorporated community for which the U.S. Census Bureau doesn't even keep official population estimates."

Democratic Rep. Dennis Cardoza, no doubt frustrated the railroad money wasn't going into his district, dubbed it the "train to nowhere." In reality, "tracks to nowhere" might have been a bit more accurate, as the Authority never had plans to run actual high-speed trains until more than 200 miles of the system are built, and the rails reach one of California's major cities.

The *Mercury News* reported, "Even rail authority member Rod Diridon, of San Jose, said he spent four hours on the phone with authority staffers trying to make sense of it. 'I'm still struggling to understand why the originating system wouldn't interconnect to major communities,' Diridon said." Diridon, according to the article, went on to say that "federal officials have threatened to yank funds if the authority hasn't chosen a starting point by the end of the month . . . "

As the Authority was fighting off the "train to nowhere" label, it had some good luck—incoming Republican governors John Kasich and Scott Walker turned down high-speed rail stimulus funds granted to Ohio and Wisconsin, saying they couldn't afford the rail projects given their states fiscal problems. The Obama administration acted quickly to divert $624 million of the rejected money to the California project. The new infusion, with matching amounts from California, allowed the Authority nearly to double the proposed first segment, to 123 miles, and make it reach just north of Bakersfield (though short of its downtown).

Officially, the Authority gives several explanations for the location of the first section. The cost of building in the Central Valley and of acquiring any land is cheaper, and the terrain is much flatter.

Rachel Wall, a spokeswoman for the Authority, explains that "that is the most amount of infrastructure for the least amount of money."

In addition, under the terms of the federal funding, California must have a contingency plan for using the new tracks if the whole system doesn't get built. The current plan would connect up with existing tracks, allowing Amtrak to use them if all else fails. Of course, the $5.5 billion cost of the first phase of construction is a high price for a set of tracks that would merely enable Amtrak to slightly increase speeds for that one stretch.

"Our intention is to build the statewide system," Wall emphasizes. "That is our sole intention."

The centerpiece of the proposed 800-mile system is a line that promises to transport passengers from San Francisco to Los Angeles in two hours and 40 minutes, with extensions planned to Sacramento and San Diego.

The major dilemma in rail construction is that building in the higher-density population areas where there is more demand for public transportation creates more resistance. Yet by choosing to start in the middle of the state, the Authority is gambling that it will be able to overcome community opposition as it approaches the major population centers.

The project is already encountering opposition from the agricultural community in the supposedly easy section of the line. Diana Peck serves as the executive director of the Kings County Farm Bureau, based in Hanford, which is near the midpoint of the proposed first segment. "It's not the path of least resistance that they think it is," she says of the Authority's thinking. "It's disheartening when people don't understand the magnitude of agricultural production and what it requires, and they think they're just going to come and cut through open space. It's not the case."

What worries farmers is that the proposed route would divide their land at a diagonal, disrupting their expensive irrigation systems. Given that the railroad would need at least 100 feet of right of way, farmers don't know how they'll transmit water to their land that would now be on the other side of the tracks. Even if they can figure that out, given that high-speed trains will be whizzing by at 220 miles per hour, there won't be the normal number of railroad crossings. Farmers fear that they'll have to drive six to eight miles just to get back and forth on their own property. Even worse, some parcels of land could be impossible to get to at all, boxed in between the tracks and somebody else's private property. Such complications would depress the value of the farms in a region where agriculture is the leading industry.

Peck says that farmers have raised their concerns to rail officials on many occasions, even traveling to meet with them in Sacramento. But she laments that while the officials listen, they have not actually responded to questions. The option of legal action remains on the table if the Authority continues to ignore them, she says.

The Authority will also run into complaints as it moves north to Madera, where officials are already raising concerns over one of the proposed routes. "We see more problems and issues than any benefit," Madera mayor Robert Poythress says.

One of the concerns is that the town, which is already divided by a highway, would become further divided by the high-speed rail tracks. Local officials are pushing for the Authority to build a park along the tracks, or do something else to mitigate the effects of having bullet trains shoot through their town, yet the budget is unlikely to allow for that. "Right now, we've been only promised negative impacts," Poythress said. "I think they're just staying, 'Damn the torpedoes, full speed ahead,' that once they build it people will see how neat it is and resistance will subside."

Azteca Milling, which processes corn for making tortillas and tamales, has said it would leave California if the Authority chooses a proposed route that would force them to give up their Madera plant. The plant has more than 100 employees, and has corn contracts with 45 local farmers, according to the Madera County Economic Development Commission.

But such resistance is minor compared to what the Authority is already encountering along the San Francisco Bay coast, where several affluent communities are up in arms over a proposed route that they fear would tear apart their downtowns. The cities of Palo Alto, Atherton, and Menlo Park have joined several environmental groups in a lawsuit against the Authority, hoping to stop that route by challenging the environmental review process.

"The problem with this routing is it runs literally right through the downtowns of these cities," said Stuart Flashman, an Oakland-based attorney representing the plaintiffs. "Back in the early 20th century, there were all these elevated railroads being built in the middle of cities, and what we learned is they ended up causing a lot of blight. And these cities realize that if you put this 40-foot-high elevated rail structure through our downtown areas, we're going to have blight."

While the cities would be willing to accept a tunnel instead of an elevated track, Authority officials have thus far balked at the cost—and rail supporters have argued that if those cities want a tunnel, they should pay for the added cost, not taxpayers in the rest of the state.

Some of Flashman's clients are environmental groups who are predisposed to liking high-speed rail, but are concerned that as the train spreads out into rural areas and spurs growth, it could lead to sprawl.

"What my clients who support high-speed rail are concerned about is that the result of this project is going to be a poorly done project that will sour Californians, and perhaps the whole country, on high-speed rail," he said. "If it's done poorly, it will leave a sour taste in people's mouths, like redevelopment in the 1950s, where people came in and said, 'we're going to do slum clearance' and they tore down whole communities. And today, people cringe whenever anybody mentions redevelopment, because people remember that. And that's what they're worried about, that this will leave a lot of bad memories that will poison the waters for years to come."

In a series of court decisions in 2009, a judge ruled that the Authority had to redo its environmental impact report, but denied a request to halt further work on the project. The Authority has argued that the route that is preferred by the cities would increase travel times and require the taking of more property through eminent domain.

Wall, the Authority spokeswoman, says they are continually trying to work closely with local communities to "minimize impact as much as possible."

"Communities throughout the state have concerns about the project and they have a right to, and that's why we've set such a vigorous environmental review process," she says. "We have an overwhelming support by Californians to do something about our transportation alternatives here. So there still is support for high-speed rail in California."

Even if California rail builders are able to overcome community opposition, the Authority has been subject to critical reports in the past year from the state auditor, inspector general, legislative analyst's office, an independent review group, and the University of California at Berkeley, raising questions about either planning, the business model, or the ridership estimates.

Under the terms of the ballot measure, the high-speed rail system, once built, cannot have its operating costs subsidized by taxpayers, an unusual arrangement for public transportation projects. Measuring demand is obviously the key ingredient to making business projections, but predictions for ridership have been wildly at odds. In 2007, the Authority commissioned a study that estimated that the high-speed rail system would attract more than 100 million passengers by 2030. Two years later, the number had dropped dramatically, to 39 million. And according to Berkeley's Institute of Transportation Studies, that still isn't a reliable figure.

"We found that the model that the rail authority relied upon to create average ridership projections was flawed at key decision-making junctures," said the project's main researcher, Samer Madanat, professor of civil and environmental engineering, in a press statement announcing the June 2010 study. "This means that the forecast of ridership is unlikely to be very close to the ridership that would actually materialize if the system were built. As such, it is not possible to predict whether the proposed high-speed rail system will experience healthy profits or severe revenue shortfalls."

One of the problems with the model that the Authority relied on was that the individuals surveyed were not representative of the typical California traveler. "For example," the study notes, "nearly 90% of long distance (more than 100-mile) business passenger trips are made by car, while 78% of the long distance business travelers sampled for the study were traveling by air."

Price will be a key determinant of demand, and yet those estimates have also varied wildly. The Authority initially claimed that a one-way ticket between Los Angeles and San Francisco would cost $55, but that price has already nearly doubled, to $105.

Wall responds that the initial ridership models, done by the respected firm Cambridge Systematics, were conducted for the narrower purpose of assessing environmental impact, and they shouldn't be viewed as the final numbers. "The ridership model will continue to be refined as the project continues, as the environmental [review] process continues," she says. "So it wasn't a one-shot chance at the ridership model."

She also cited a survey by the American Public Transportation Association finding that 62 percent of Americans said they would "definitely" or "probably" use high-speed rail if it were a travel option.

However, as long as there is uncertainty about the ridership potential, it adds to the other problems facing the project. Doubts about ridership fuel community opposition, because people ask why they should have to put up with disruption to build a rail service that won't be utilized. And in the coming years, it will be harder to convince a more skeptical Congress—not to mention private companies—to invest in a project where it's unclear that enough demand exists to justify the cost.

In December, California's projected budget deficit swelled to $25.4 billion, raising further doubts about whether the state could afford such an extravagant public works project

If the system is fully built, it would cost an estimated $647 million a year for 30 years for state taxpayers to pay off the $9.9 billion in bonds that would need to be issued. State Sen. LaMalfa says that this is particularly irresponsible at a time when people are already protesting proposed cuts to education and health care services.

"We do not have the money to make the payments on this rail system," LaMalfa explains.

Madera's Poythress echoes similar sentiments. "I have to salute the states that turned down the money," he says, referring to Wisconsin and Ohio, "because I think what they're doing is turning down future problems. But we seem to be welcoming the money with open arms."

The only way that the Authority is able to unlock the bond money is to receive funds from other sources. So far, that has meant federal funding, as the private sector sits on the sidelines.

"We know that private funding will probably materialize after the big federal funding commitments come through," Wall acknowledges. Some have raised questions about the reliability of the estimate that the entire system would cost $43 billion. A Reason Foundation study projected that the real cost would be far higher, in the range of $65 billion to $81 billion. Wendell Cox, one of the authors of the study, says that based on international data, the cost of these types of projects tends to be underestimated by 40 percent to 100 percent, because public officials have every incentive to lowball the figures to get projects approved.

Even if the $43 billion is accepted at face value, the funding is still a daunting challenge. According to the Authority's business plan, it needs to receive at least $18 billion in federal funding to complete the project, with the rest coming from the state bonds and private sources. Yet the project received a windfall in 2010 largely due to a confluence of historical factors that are unlikely to be repeated, namely, a Democratic president and Congress coming into power with a crisis they didn't want to waste.

"We got the most out of any state in the nation, $3.2 billion," Wall boasts. "So when we matched that with the state funding, we have $5.5 billion after one year of pursuing federal funding heavily."

With the change of leadership in Congress, there are several proposals by members of the new Republican majority in the House aimed at rescinding unspent stimulus funds and denying future funding to high-speed rail. While it's a matter of some debate on Capitol Hill as to whether money that's already been designated for the California high-speed rail project can be canceled, the state is clearly facing a less sympathetic federal financing environment this year.

The California High-Speed Rail Peer Review Group, an independent panel created to oversee the project, raised alarms in November 2010 over the Authority's inadequate staffing levels, business model, risk management, and financial uncertainty. On the latter point, the group's report wrote: "In light of the public concern over excessive government spending, how will the Authority close the gap between any funding resources and the project's total estimated cost? What will the Authority's course of action be if the funding gap cannot be closed?"

The Authority is undeterred, and Wall says it will continue to "compete aggressively for federal funding." Ideally, the Authority would like to see an ongoing appropriation of more than $1 billion a year for the project, a commitment that its officials claim will attract private investment.

LaMalfa has introduced a bill in the California senate that would halt any spending on the high-speed rail project until a new, independent analysis could be conducted on the projected cost. After that, he would support holding a new ballot referendum on high-speed rail, providing voters with more accurate data than existed when it went before voters in 2008. In the meantime, he's urging members of Congress to do what they can to rescind funds directed to the high-speed rail project, and deny any future requests.

"The money the federal government will be sending us will cost us money in the state, because it props up something that doesn't add up," LaMalfa says. "If they [cancel the funding], I think the whole house of cards comes down no matter what we do here in the [California] legislature."

4

—

The Grid:
Power and Communications Networks

Courtesy of Tim Boyle/Getty Images

A sign that reads "Electric Wires DANGER High Voltage" is displayed on steel towers carrying power transmission lines on August 18, 2003, in Mount Prospect, Illinois.

Giant wind turbines are powered by strong prevailing winds on May 13, 2008, near Palm Springs, California.

Editor's Introduction

The National Academy of Engineering dubbed the North American power grid "the greatest engineering achievement of the last century," Joel Achenbach writes in this chapter's first selection, "The 21st Century Grid: Can We Fix the Infrastructure that Powers Our Lives?" The vast network has also been called "the world's biggest machine," according to Achenbach. Despite such hyperbole, the grid is not without its flaws, many of which are likely to become more obvious in the years ahead. One of those shortcomings is hinted at in the first quote. The grid is not getting any younger. At the same time, a growing population is asking more and more of it.

The articles in this chapter examine the health of our power and communications networks. Among the issues addressed are the updating and "greening" of the grid: how to adapt our infrastructure to support an expanding and increasingly technology-dependent society while also transitioning from fossil-fuel-based energy to more ecologically sound alternatives.

Achenbach offers a brief history lesson in his piece, charting the grid's origins in the 19th century, when the renowned inventor Thomas Alva Edison constructed the first power plant in Lower Manhattan, and its development over the last 130 years. He discusses the need for a "smarter grid," one that allows consumers to more closely monitor their consumption, and describes our vast power and communication network as "a kludge. . . . an awkward, inelegant contraption that somehow works."

"The grid is in trouble," Ben Prawdzik warns in the subsequent entry, "Problems Facing Our Energy Infrastructure: A Critical Look at the Deteriorating North American Electrical Grid." The difficulties confronting the grid, he contends, correspond to four major categories: consumer choice, efficiency and environmental impact, security, and reliability and affordability. While a consensus has yet to emerge on what the precise solutions are, Prawdzik observes, "any effort to solve America's energy crisis will require a joint, concerted effort of government, businesses, and households."

A notorious blackout in the summer of 2003 provides the opening for "Are You Afraid of the Dark?" Jared Wade's assessment of the American electrical grid. One of the issues he identifies is the question of regulation versus deregulation in the power and communications industries. In the United States, in Wade's view, the loosening of government rules has not improved the system. While it did open up the sector to additional companies, the increased competition resulted in smaller profit margins and thus less money with which to update the infrastructure. According to Wade, "things will not really begin to improve until the federal government takes a more defined role in managing the national grid."

Lynne Kiesling also tackles the question of deregulation in the subsequent selection, "Electric Intelligence: Establishing a Smart Grid Requires Regulatory Reform, Not Subsidies." Unlike Wade, she sees deregulation, especially at the state level, as essential to building the grid of the future. The grid she envisions is one in which the recent revolution in digital technology is fully incorporated into the electricity industry. "Imagine a future," she writes, "in which your home has a system that connects all its appliances, entertainment systems, [etc.] into one communication network." Unfortunately, Kiesling believes, the current regulatory environment makes realizing this vision impossible. "All that stands in the way of vibrant customer-friendly electricity products and services," she writes, "is an outdated infrastructure run by hesitant monopolies and regulated by bureaucrats with little incentive to improve things. We can do smarter."

The Future Renewable Electric Energy Delivery and Management Systems (FREEDM) Center is at the forefront of efforts to develop smart grid technologies. The federally funded FREEDM Center is an innovative public-private partnership composed of seven universities and 44 companies. At its core, the organization's mission is to develop technologies that will improve the grid and promote renewable energy production. In "Wiring the Revolution," Thomas K. Frose reports on some of the organization's initiatives.

The 21st Century Grid[*]

Can We Fix the Infrastructure that Powers Our Lives?

By Joel Achenbach
National Geographic, July 2010

We are creatures of the grid. We are embedded in it and empowered by it. The sun used to govern our lives, but now, thanks to the grid, darkness falls at our convenience. During the Depression, when power lines first electrified rural America, a farmer in Tennessee rose in church one Sunday and said—power companies love this story—"The greatest thing on earth is to have the love of God in your heart, and the next greatest thing is to have electricity in your house." He was talking about a few lightbulbs and maybe a radio. He had no idea.

Juice from the grid now penetrates every corner of our lives, and we pay no more attention to it than to the oxygen in the air. Until something goes wrong, that is, and we're suddenly in the dark, fumbling for flashlights and candles, worrying about the frozen food in what used to be called (in pre-grid days) the icebox. Or until the batteries run dry in our laptops or smart phones, and we find ourselves scouring the dusty corners of airports for an outlet, desperate for the magical power of electrons.

The grid is wondrous. And yet—in part because we've paid so little attention to it, engineers tell us—it's not the grid we need for the 21st century. It's too old. It's reliable but not reliable enough, especially in the United States, especially for our mushrooming population of finicky digital devices. Blackouts, brownouts, and other power outs cost Americans an estimated $80 billion a year. And at the same time that it needs to become more reliable, the grid needs dramatic upgrading to handle a different kind of power, a greener kind. That means, among other things, more transmission lines to carry wind power and solar power from remote places to big cities.

Most important, the grid must get smarter. The precise definition of "smart" varies from one engineer to the next. The gist is that a smart grid would be more automated and more "self-healing," and so less prone to failure. It would be more tolerant of small-scale, variable power sources such as solar panels and wind turbines, in part because it would even out fluctuations by storing energy—in the batteries of electric cars, according to one speculative vision of the future, or perhaps in giant caverns filled with compressed air.

But the first thing a smart grid will do, if we let it, is turn us into savvier consumers of electricity. We'll become aware of how much we're consuming and cut back, especially at moments of peak demand, when electricity costs most to produce. That will save us and the utilities money—and incidentally reduce pollution. In a way, we'll stop being mere passive consumers of electrons. In the 21st century we'll become active participants in the management of this vast and seemingly unknowable network that makes our civilization possible.

So maybe it's time we got to know it.

There are grids today on six continents, and someday Europe's may reach across the Mediterranean into Africa to carry solar power from the Sahara to Scandinavia. In Canada and the U.S. the grid carries a million megawatts across tens of millions of miles of wire. It has been called the world's biggest machine. The National Academy of Engineering calls it the greatest engineering achievement of the last century.

Thomas Edison, already famous for his lightbulb, organized the birth of the grid in 1881, digging up lower Manhattan to lay down copper wires inside brick tunnels. He constructed a power plant, the Pearl Street Station, in the shadow of the Brooklyn Bridge. On September 4, 1882, in the office of tycoon J. P. Morgan, Edison threw a switch. Hundreds of his bulbs lit up Drexel, Morgan & Co. and other offices nearby.

Edison was heavily invested in direct current, which worked well in his bulbs and which at the time was low voltage. Alternating current, he argued colorfully, was more appropriate to executing criminals. (He had a circus elephant electrocuted to prove his point.) The argument was misleading: AC, in which the electrons don't stream in one direction but oscillate back and forth at a given frequency, isn't intrinsically more dangerous than DC. High voltage is what's dangerous—but it's also what allows power to be transmitted hundreds of miles without excessive loss. AC won out over DC largely because it can easily be stepped up with transformers, transmitted, then stepped down again to a safer household voltage of 110 or 220. By the 1890s AC lines were running from the new Niagara Falls generating station to Buffalo, some 20 miles away. These days, ironically, high-voltage DC is sometimes preferred for very long distances; it's harder to produce than AC, but it loses even less power.

It took decades for electricity to expand from factories and mansions into the homes of the middle class. In 1920 electricity still accounted for less than 10 percent of the U.S. energy supply. But inexorably it infiltrated everyday life. Unlike coal, oil, or gas, electricity is clean at the point of use. There is no noise, except perhaps a faint hum, no odor, and no soot on the walls. When you switch on an electric lamp, you don't think of the huge, sprawling power plant that's generating the electricity (noisily, odoriferously, sootily) many miles away. Refrigerators replaced iceboxes, air conditioners replaced heat prostration, and in 1956 the electric can opener completed our emergence from the dark ages. Today about 40 percent of the energy we use goes into making electricity.

At first, utilities were local operations that ran the generating plant and the distribution. A patchwork of mini-grids formed across the U.S. In time the utilities realized they could improve reliability and achieve economies of scale by linking their transmission networks. After the massive Northeast blackout of 1965, much of the control of the grid shifted to regional operators spanning many states. Yet today there is still no single grid in the U.S.; there are three nearly independent ones—the Eastern, Western, and Texas Interconnections.

They function with antiquated technology. The parts of the grid you come into contact with are symptomatic. How does the power company measure your electricity usage? With a meter reader—a human being who goes to your home or business and reads the dials on a meter. How does the power company learn that you've lost power? When you call on the phone. In general, utilities don't have enough instantaneous information on the flow of current through their lines—many of those lines don't carry any data—and people and slow mechanical switches are too involved in controlling that flow.

"The electrical grid is still basically 1960s technology," says physicist Phillip F. Schewe, author of *The Grid*. "The Internet has passed it by. The meter on the side of your house is 1920s technology." Sometimes that quaintness becomes a problem. On the grid these days, things can go bad very fast.

When you flip a light switch, the electricity that zips into the bulb was created just a fraction of a second earlier, many miles away. Where it was made, you can't know, because hundreds of power plants spread over many states are all pouring their output into the same communal grid. Electricity can't be stored on a large scale with today's technology; it has to be used instantly. At each instant there has to be a precise balance between generation and demand over the whole grid. In control rooms around the grid, engineers constantly monitor the flow of electricity, trying to keep voltage and frequency steady and to avoid surges that could damage both their customers' equipment and their own.

When I flip a switch at my house in Washington, D.C., I'm dipping into a giant pool of electricity called the PJM Interconnection. PJM is one of several regional operators that make up the Eastern grid; it covers the District of Columbia and 13 states, from the Mississippi River east to New Jersey and all the way down to the Outer Banks of North Carolina. It's an electricity market that keeps supply and demand almost perfectly matched—every day, every minute, every fraction of a second—among hundreds of producers and distributors and 51 million people, via 56,350 miles of high-voltage transmission lines.

One of PJM's new control centers is an hour north of Philadelphia. Last February I went to visit it with Ray E. Dotter, a company spokesman. Along the way Dotter identified the power lines we passed under. There was a pair of 500-kilovolt lines linking the Limerick nuclear plant with the Whitpain substation. Then a

230-kilovolt line. Then another. Burying the ungainly lines is prohibitively expensive except in dense cities. "There's a need to build new lines," Dotter said. "But no matter where you propose them, people don't want them."

Dotter pulled off the turnpike in the middle of nowhere. A communications tower poked above the treetops. We drove onto a compound surrounded by a security fence. Soon we were in the bunker, built by AT&T during the Cold War to withstand anything but a direct nuclear hit and recently purchased by PJM to serve as its new nerve center.

In the windowless control room, dominated by a curved wall of 36 computer screens, dispatch general manager Mike Bryson explained what I was seeing. A dynamic map on one of the screens showed the PJM part of the grid. Arrows represented major transmission lines, each with a number showing how much juice was on the line at that moment. Most of the arrows pointed west to east: In the eastern U.S. electricity flows from major power plants in the heartland toward huge clusters of consumers along the eastern seaboard. At that moment PJM lines were carrying 88,187 megawatts. "Today is a mild winter day—I don't think we'll have over 90,000," Bryson said.

The computers take data from 65,000 points on the system, he explained. They track the thermal condition of the wires; too much power flowing through a line can overheat it, causing the line to expand and sag dangerously. PJM engineers try to keep the current alternating at a frequency of precisely 60 hertz. As demand increases, the frequency drops, and if it drops below 59.95 hertz, PJM sends a message to power plants asking for more output. If the frequency increases above 60.05 hertz, they ask the plants to reduce output. It sounds simple, but keeping your balance on a tightrope might sound simple too until you try it. In the case of the grid, small events not under the control of the operators can quickly knock down the whole system.

Which brings us to August 14, 2003. Most of PJM's network escaped the disaster, which started near Cleveland. The day was hot; the air conditioners were humming. Shortly after 1 p.m EDT, grid operators at First Energy, the regional utility, called power plants to plead for more volts. At 1:36 p.m. on the shore of Lake Erie, a power station whose operator had just promised to "push it to my max max" responded by crashing. Electricity surged into northern Ohio from elsewhere to take up the slack.

At 3:05 a 345-kilovolt transmission line near the town of Walton Hills picked that moment to short out on a tree that hadn't been trimmed. That failure diverted electricity onto other lines, overloading and overheating them. One by one, like firecrackers, those lines sagged, touched trees, and short-circuited.

Grid operators have a term for this: "cascading failures." The First Energy operators couldn't see the cascade coming because an alarm system had also failed. At 4:06 a final line failure sent the cascade to the East Coast. With no place to park their electricity, 265 power plants shut down. The largest blackout in North American history descended on 50 million people in eight states and Ontario.

At the Consolidated Edison control center in lower Manhattan, operators remember that afternoon well. Normally the power load there dips gradually, minute by minute, as workers in the city turn off their lights and computers and head home. Instead, at 4? p.m. lights went out in the control room itself. The operators thought: 9/11. Then the phone rang, and it was the New York Stock Exchange. "What's going on?" someone asked. The operators knew at once that the outage was citywide.

There was no stock trading then, no banking, and no manufacturing; restaurants closed, workers were idled, and everyone just sat on the stoops of their apartment buildings. It took a day and a half to get power back, one feeder and substation at a time. The blackout cost six billion dollars. It also alarmed Pentagon and Homeland Security officials. They fear the grid is indeed vulnerable to terrorist attack, not just to untrimmed trees.

The blackout and global warming have provided a strong impetus for grid reform. The federal government is spending money on the grid—the economic-stimulus package allocated $4.5 billion to smart grid projects and another six billion dollars or so to new transmission lines. Nearly all the major utilities have smart grid efforts of their own.

A smarter grid would help prevent blackouts in two ways. Faster, more detailed feedback on the status of the grid would help operators stay ahead of a failure cascade. Supply and demand would also be easier to balance, because controllers would be able to tinker with both. "The way we designed and built the power system over the last hundred years—basically the way Edison and Westinghouse designed it—we create the supply side," says Steve Hauser of the U.S. Department of Energy's National Renewable Energy Laboratory (NREL) near Boulder, Colorado. "We do very little to control demand."

Working with the NREL, Xcel Energy has brought smart grid technology to Boulder. The first step is the installation of smart meters that transmit data over fiber-optic cable (it could also be done wirelessly) to the power company. Those meters allow consumers to see what electricity really costs at different times of day; it costs more to generate during times of peak load, because the utilities have to crank up auxiliary generators that aren't as efficient as the huge ones they run 24/7.

When consumers are given a price difference, they can choose to use less of the expensive electricity and more of the cheap kind. They can run clothes dryers and dishwashers at night, for instance. The next step is to let grid operators choose. Instead of only increasing electricity supply to meet demand, the operators could also reduce demand. On sweltering summer days the smart grid could automatically turn up thermostats and refrigerators a bit—with the prior agreement of the homeowners of course.

"Demand management" saves energy, but it could also help the grid handle renewable energy sources. One of the biggest problems with renewables like solar and wind power is that they're intermittent. They're not always available when demand peaks. Reducing the peak alleviates that problem. You can even imagine programming smart appliances to operate only when solar or wind power is available.

Some countries, such as Italy and Sweden, are ahead of the U.S. in upgrading their electrical intelligence. The Boulder project went online earlier this year, but only about 10 percent of U.S. customers have even the most primitive of smart meters, Hauser estimates.

"It's expensive," he says. "Utilities are used to spending 40 bucks on an old mechanical meter that's got spinning dials. A smart meter with a software chip, plus the wireless communication, might cost $200—five times as much. For utilities, that's huge." The Boulder project has cost Xcel Energy nearly three times what it expected. Earlier this year the utility raised rates to try to recoup some of those costs.

Although everyone acknowledges the need for a better, smarter, cleaner grid, the paramount goal of the utility industry continues to be cheap electricity. In the U.S. about half of it comes from burning coal. Coal-powered generators produce a third of the mercury emissions in America, a third of our smog, two-thirds of our sulfur dioxide, and nearly a third of our planet-warming carbon dioxide—around 2.5 billion metric tons a year, by the most recent estimate.

Not counting hydroelectric plants, only about 3 percent of American electricity comes from renewable energy. The main reason is that coal-fired electricity costs a few cents a kilowatt-hour, and renewables cost substantially more. Generally they're competitive only with the help of government regulations or tax incentives. Utility executives are a conservative bunch. Their job is to keep the lights on. Radical change makes them nervous; things they can't control, such as government policies, make them nervous. "They tend to like stable environments," says Ted Craver, head of Edison International, a utility conglomerate, "because they tend to make very large capital investments and eat that cooking for 30 or 40 or 50 years."

So windmills worry them. A utility executive might look at one and think: What if the wind doesn't blow? Or look at solar panels and think: What if it gets cloudy? A smart grid alone can't solve the intermittence problem. The ultimate solution is finding ways to store large amounts of electricity for a rainy, windless day.

Actually the U.S. can already store around 2 percent of its summer power output—and Europe even more—behind hydroelectric dams. At night, when electricity is cheaper, some utilities use it to pump water back uphill into their reservoirs, essentially storing electricity for the next day. A small power plant in Alabama does something similar; it pumps air into an underground cavern at night, compressing it to more than a thousand pounds per square inch. During the day the compressed air comes rushing out and spins a turbine. In the past year the Department of Energy has awarded stimulus money to several utilities for compressed-air projects. One project in Iowa would use wind energy to compress the air.

Another way to store electricity, of course, is in batteries. For the moment, it makes sense on a large scale only in extreme situations. For example, the remote city of Fairbanks, Alaska, relies on a huge nickel-cadmium, emergency-backup battery. It's the size of a football field.

Lithium-ion batteries have more long-term potential—especially the ones in electric or plug-in-hybrid cars. PJM is already paying researchers at the University of Delaware $200 a month to store juice in three electric Toyotas as a test of the idea. The cars draw energy from the grid when they're charging, but when PJM needs electricity to keep its frequency stable, the cars are plugged in to give some back. Many thousands of cars, the researchers say, could someday function as a kind of collective battery for the entire grid. They would draw electricity when wind and solar plants are generating, and then feed some back when the wind dies down or night falls or the sun goes behind clouds. The Boulder smart grid is designed to allow such two-way flow.

To accommodate green energy, the grid needs not only more storage but more high-voltage power lines. There aren't enough running to the places where it's easy to generate the energy. To connect wind farms in Kern County with the Los Angeles area, Southern California Edison, a subsidiary of Edison International, is building 250 miles of them, known as the Tehachapi Renewable Transmission Project. A California law requires utilities to generate at least 20 percent of their electricity from renewable sources as of this year.

Green energy would also get a boost if there were more and bigger connections between the three quasi-independent grids in the U.S. West Texas is a Saudi Arabia of wind, but the Texas Interconnection by itself can't handle all that energy. A proposed project called the Tres Amigas Superstation would allow Texas wind—and Arizona sun—to supply Chicago or Los Angeles. Near Clovis, New Mexico, where the three interconnections already nearly touch, they would be joined together by a loop of five-gigawatt-capacity superconducting cable. The three grids would become, in effect, one single grid, national and almost rational.

Studying the map of the grid in the PJM control room, I noticed unfamiliar place-names: Amos, Pruntytown, Matt Funk, Sporn. Washington, D.C., was not labeled; Mike Bryson suggested it was somewhere near a substation called Waugh Chapel. One of the largest generating stations on the map, he added helpfully, was the Gavin plant, which at that moment was cranking out 2,633 megawatts.

Where's Gavin? I asked.

"West Virginia or Kentucky somewhere," Bryson said.

It's actually in southern Ohio. The grid is a kind of parallel world that props up our familiar one but doesn't map onto it perfectly. It's a human construction that has grown organically, like a city or a government—what technical people call a kludge. A kludge is an awkward, inelegant contraption that somehow works. The U.S. grid works well by most measures, most of the time; electricity is abundant and cheap.

It's just that our measures have changed, and so the grid must too. The power industry, says Ted Craver of Edison International, faces "more change in the next ten years than we've seen in the last hundred." But at least now the rest of us are starting to pay attention.

Problems Facing Our Energy Infrastructure[*]

A Critical Look at the Deteriorating North American Electrical Grid

By Ben Prawdzik
Yale Economic Review, Winter/Spring 2011

What was the most important technological development in the last 100 years for the advancement of human civilization and society? Perhaps the microchip, penicillin, the Internet, or DNA sequencing? According to the United States National Academy of Engineering, the answer is North America's electric grid. Experts at the Academy report that "widespread electrification gave us power for our cities, factories, farms, and homes—and forever changed our lives . . . from street lights to supercomputers, electric power makes our lives safer, healthier and more convenient."

One might be surprised at the Academy's finding; most are unaware of the full technological complexity behind America's mammoth power generation and transmission operation. But though most of us may not realize it, the grid touches on every facet of our lives each day. From our cell phones to our appliances, the grid is important. And more importantly, the grid is in trouble.

A product of 1950's technological design, the grid is a rapidly aging, antiquated machine. Domestic energy demand has far outstripped investments in increased capacity over the past twenty-five years, and with growing electricity needs predicted for the future, our attitude towards the electric grid must change.

THE CRISIS: PROBLEMS FACING OUR ENERGY INFRASTRUCTURE

The North American electric grid faces a spectrum of important, fundamental challenges in the near and long term future. Twentieth century designs and dated technology in a modern, computerized era have put the grid at its capacity limit.

Its problems can be summarized by four primary categories: reliability and afford-ability, efficiency and environmental impact, security, and consumer choice.

RELIABILITY AND AFFORDABILITY

The most pressing challenge facing America's electric grid is the question of reli-ability and affordability. The issue stems from one, simple fact: America's energy demand has been expanding exponentially over the past 25 years while invest-ments in capacity and transmission technology have declined. Since the 1990s, domestic electricity demand has risen over 25 percent while [...] construction of transmission facilities has decreased by about 30 percent. Furthermore, NREL estimates that, through the implementation of energy efficiency programs and in-creased use of renewables, we have the potential not only to displace that growth in carbon emissions, but to actually reduce our domestic carbon output to below 1,000 million tons by 2030. The figures speak for themselves: we simply cannot afford to ignore the efficiency issue any longer.

SECURITY

Electricity is the lifeblood of our society, powering everything from households to hospitals. Our reliance on the grid becomes clear with each blackout, when our way of life comes to a screeching halt without power. Thus, it is critical that the electric grid remain safe and operational. However, the growing reliance of utilities on internet-based communication has increased the vulnerability of grid control systems.

Our grid's vulnerability is clearly exposed when we consider the potential dan-gers from cyberspace. In April 2009, spies traced to China and Russia penetrated the U.S. electrical grid system and left behind software packages capable of de-stroying system components. Intelligence officials are aware that the Chinese and Russian governments have attempted several times to map our infrastructure, and, according to the Department of Homeland Security, the number of intrusions into the system is growing. Given the functional importance of the electric grid in our society, these cyber attacks pose a major risk to national security. Spies and hackers can operate from almost anywhere in the world with nearly complete anonymity, and hacking is cheap enough to be a potential tool for a significant terrorist attack. The costs of cyber attacks are staggering—in just the past 6 months, the Obama administration has spent over $100 million repairing cyber damage to our electric grid. A large-scale attack would have the potential to paralyze the economy.

CONSUMER CHOICE

Our current energy grid places immense constraints on consumer choice. Consumers in the 21st century have evolved complex lifestyles, tastes, preferences, and living habits. We live in an age in which we have come to expect access to more information in less time and with greater ease of use. With more information, we try to make more informed decisions. Accordingly, the way we think about spending money is changing, and electricity is long overdue to adapt to our new lifestyles.

With our current electric infrastructure, we are locked into antiquated pricing models for energy. Consumers do not have the option to avoid peak hour cost increases and adapt more efficient lifestyles based on energy pricing. Furthermore, our grid is ill-equipped to handle what many anticipate will be a wave in "green" technology implementation. Experts warn that the grid, as it stands, will not be able to handle the large-scale use of plug-in hybrid electric vehicles, as the massive spike in energy demand will overwhelm urban transmission lines. Households with solar panels are generally barred from transferring excess capacity back into the grid, and the technology in place makes it nearly impossible logistically to incorporate the thousands of new renewable power generation stations we need to build. In summary, the technology behind our electric grid has become outdated, and if we hope to expand reliability, affordability, efficiency, security, and choice, we need to make a renewed, concerted investment in our grid.

SOLUTIONS TO THE CHALLENGES FACING OUR ENERGY FUTURE

We know that much of the electric grid will need to be replaced within the next thirty years. Therefore, it is important to put policies in place now that will set the grid's future on the right track. There is no single, "silver bullet" solution with which to addressing the challenges facing the energy grid in the coming decades; rather, the U.S's approach needs to be more like a "shotgun slug"—a number of coordinated reforms, incentives, and investments to change our attitude towards energy and create market conditions that will facilitate socially optimal investment decisions. These policy changes can be targeted to three main areas of grid reform: upgrading the physical transmission and distribution systems, expanding power generation infrastructure, and aggressively promoting energy efficiency. The following provides an outline of several promising solutions addressing each of the three principles above. While the following outline certainly does not encompass every solution, it nonetheless provides an insightful look at the kinds of approaches needed to ensure America's energy future.

UPGRADE PHYSICAL TRANSMISSION AND DISTRIBUTION INFRASTRUCTURE

Investing in upgrades to America's energy grid transmission and distribution infrastructure is arguably the most important step in addressing the issues facing the grid, because the right improvements open the grid up to a number of other powerful improvements. Many experts agree that we need to pursue what has become known as a "smart grid." In today's system, utilities need to predict exactly how much electricity will be needed at different times throughout the day, and power plants must either ramp up or slow down generation to conform to these predictions. These calculations require an incredible amount of computation and manpower, and any blip in the system—any transmission line that dies or plant that goes offline—can be tremendously complex to rectify.

A smart grid would use digital technology to collect consumption, generation, and transmission data in real time and incorporate this data in making automated decisions as to how to most efficiently manage the grid infrastructure. This would require the installation of sensors at power plants and along transmission lines, as well as smart meters—consumer end, digital readers that relay information about energy pricing, capacity, and demand between the utility companies and the end user.

The economic benefits of creating a smart grid infrastructure would be enormous. With automated technology, previously intensive work such as determining which transmission lines to use or locating breaks in the infrastructure could be done by computer, making the grid more efficient and reliable. According to the Electricity Sector Framework for the Future, these changes could result in $1.8 trillion of annual additive revenue by 2020, and power disturbance costs to the U.S. economy could drop by as much as $49 billion per year. A flexible, dynamic smart grid would also give consumers more decision making ability, as they would be able to see the cost of peak hour use through the use of smart meters and reduce their energy consumption at those times to save cash. A smart grid would also be able to accommodate the integration of decentralized renewable sources such as local wind turbines and solar arrays. Widespread use of PHEVs would be feasible with automated grid management as well. Finally, smart grids, if designed properly, would allow for added security against external cyberspace threats.

EXPAND POWER GENERATION INFRASTRUCTURE

In addition to upgrading transmission and distribution systems, we need to ensure that future investments in power generation infrastructure work to achieve two major principles: expanded use of renewable energy and decentralization of power generation. One interesting solution with which to expand the U.S. electric grid's power generation capacity would be to foster the a widespread development of "community wind" projects. "Community wind" refers to a class of wind energy ownership structures through which wind developments are at least partially owned by individuals or businesses in the local area surrounding the project. Community ownership provides a new source of income for the area as newly created construction and maintenance jobs bring additional wages into the community. By letting local citizens tap into the wealth generated by wind energy, communities will be more open to the idea of wind as a source of energy. And because the electricity would be generated locally, there would be no need for expensive, leaky long distance transmission lines. Finally, community wind projects would help to develop electric infrastructure in new areas of the country, thereby decentralizing our power grid and increasing energy security.

AGGRESSIVELY PROMOTE ENERGY EFFICIENCY

The third and final component to solving America's energy crisis is to aggressively promote efficient energy usage on the part of the consumer. Standards for conventional household appliances and home electric wiring can be established to ensure greater efficiency in energy use. State governments can begin to promote "smart" building codes, mandating higher standards for thermal insulation and wider use of efficient building materials to decrease energy demand. Furthermore, a building code that requires the implementation of smart meters ensures that future construction in America can be easily integrated into an evolving energy grid.

Finally, the government needs to change the way in which we address externalities. The release of greenhouse gasses by businesses affects human health and the environment—huge external costs that can be completely disregarded by firms. Whether through a carbon tax, cap and trade policy, or some other legislative measure, the government needs to work to internalize these costs.

CONCLUSION

It is clear that America's energy grid faces a number of critical challenges in the future. From issues of capacity, reliability, efficiency, security, and choice, our aging grid is in dire need of repair. And while the specific policy choices we should follow to address the issues ahead are still up for debate, it is clear that any effort to solve America's energy crisis will require a joint concerted effort of government, businesses, and households. But above all else, one thing is clear: whether we realize it or not, the electric grid is one of the fundamental cornerstones of modern society. Our living styles are changing, and to support those changes, we need to change the way we harness and manage electricity. It is our duty to, inform ourselves about the grid and support the investments we need in the system now—America's energy future depends on it.

Are You Afraid of the Dark?*

By Jared Wade
Risk Management, May 2004

On August 14, 2003, the largest blackout in North American history left more than 50 million people in the dark across nine states in the United States and many parts of Canada. In addition to the logistical and social problems that ensued, the economic fallout was tremendous.

Businesses in New York alone are estimated to have lost in excess of $800 million, mostly due [to] lost productivity. According to a study conducted jointly by business consulting firm Miriflex and Case Western University, more than two-thirds of surveyed executives affected by the Northeast blackout responded that their company lost a full business day. One-fourth said that each hour of downtime cost their company over $50,000, which for those who lost a full day, was a loss of more than $400,000.

While the Northeast blackout was devastating, power concerns are not isolated to the United States. As technology advances and the demand for energy increases globally, every power system in the world is feeling the strain.

For instance, shortly after the Northeast blackout, Italian Prime Minister Silvio Berlusconi publically insisted that a disaster of this magnitude could never happen in Italy, which receives nearly 20 percent of its energy from other European nations. Weeks later, after multiple transmission lines from other countries failed, Italy suffered a countrywide blackout that left some 57 million people without power.

Eerily, major blackouts also shut off power to London, San Francisco and parts of Scandinavia in the latter months of 2003 as well. While these three outages were much shorter and had much less economic impact, it does beg the question of how safe can any power supply be. The increasing demand for larger amounts of energy to be transmitted over longer distances will only raise the stakes of this problem.

With this in mind, companies are left to ponder whether a power outage like the one in the Northeast was a fluke occurrence or something that could happen more frequently in the future. With five major blackouts worldwide, will 2003 simply go down as the Year of the Blackout or should this be taken as a sign of things to come?

POWER STRUGGLE

While power grids around the world differ, most modern nations essentially transmit energy in the same manner. "You've basically got three points in the chain," says Paul Greenhill, partner of the risk services company, JLT Group's Energy Business Unit. "You've got the people that produce the power—the wholesale energy market, which to a certain extent operates under the watchful eye of a regulator but also operates in a competitive environment. Then you've got the national grid, which basically just moves the stuff around. The grid picks electricity up from the power stations and moves it to the local power distribution companies. Then the local distribution companies are the ones that actually take the energy off the high voltage network and deliver it at low voltage to people's homes and other end users."

In the United States, especially since the Northeast blackout, the national grid has been criticized for being outdated and unable to cope with serious outages. Very few upgrades have been made over the past 20 years and some parts of the system are operating near capacity. Some experts estimate that any investment to significantly upgrade the grid could exceed $100 billion. "We're a superpower with a third world grid," former energy secretary Bill Richardson said in a recent CNN interview. "The problem is that nobody is building enough transmission capacity."

While it is a somewhat inexact science, some statistics figure that U.S. power demand has increased by 30 percent in the last decade, while transmission capacity has grown by only 15 percent. And just like a highway without enough lanes, this leads to congestion and traffic problems.

Due to state regulations and independent power suppliers, problems also can arise from different infrastructure used in different areas. "If you've got five or six competing technologies, than it's obviously much more difficult for people to understand what's going on in a situation like that in [the Northeast]," says Greenhill. "It seemed to be that people just didn't know what was happening. People were looking at the screens and they were seeing things flicking off but they didn't know what to do about it."

In some cases, competing infrastructure is not only a regional problem but one that occurs within every state and sometimes even within individual towns. "One of the problems with the system is that you can literally have a one-horse town somewhere in the middle of Idaho that will have its own little power generation plant," says Greenhill. "This plant could have been set up by the town 70 years ago—possibly using 70-year-old technology—and this energy is sort of passed around the town on its own little circuit which then links up to the national grid in one of many ways."

Situations like this create a system that is set up like a patchwork quilt in which different circuits, production systems and transmission lines all link together without uniformity. With such a system in place it cannot be altogether surprising that disruptions will occur as competing infrastructure attempts to work together.

"Sometimes you see the power flows changing altogether," says Tim Cherry, a senior manager for PricewaterhouseCoopers' utilities practices. "When new markets are introduced, there are times when transmission that has been running from east to west has changed directions, now going west to east. This is all very regionally based. In some cases, additional power plants need to be put into the system. In other parts of the country the actual transmission system itself needs to be redesigned."

In the United States, different states can have their own infrastructures that while internally legal and compliant with regulations, may not be so in the states they exchange energy transmissions with, says Greenhill. "Electricity doesn't recognize state boundaries," says Greenhill. "So as it goes across a state it may be compliant with regulations but actually, electrically it doesn't work and there's an incompatibility with some of the systems that have grown up in the different states."

UPDATING THE SYSTEM

While the size, scope and magnitude of the North American system presents a greater challenge to creating a unified system than that faced by many European nations, the United States may want to look across the Atlantic for some possible solutions. Despite the recent London blackout, the grid system in the United Kingdom is considered to be very modern and this alone allows it to sidestep some of the problems seen in the United States.

For example, because of the U.K. system's uniformity, the London blackout lasted a mere 45 minutes and regulators were quickly able to identify the problem—which turned out to be one improperly installed circuit—and get the system back on track.

"In the United Kingdom, it's the same system everywhere, so the guys that operate it can go into an office in the south or north of the country and they'll see exactly the same system and it will be no problem for them," says Greenhill. "In the United States, which you would think would have a extremely robust grid, there's a real patchwork quilt of different things that don't really fit well together, whereas in the U.K. all the power is distributed on the same basis."

The current standing of the U.K. grid came out of efforts made during the 1980s, when the Margaret Thatcher administration moved towards privatized systems, which were believed to work better than nationalized ones. "So what they did with power was to acknowledge that you can successfully privatize power generation and power delivery, but the middle part—how you join the two together—because it's a national grid and because electricity flows around in a circuit, it makes sense to have this step run by one company that is basically an expert at running a circuit," says Greenhill. "Then theoretically, everyone has a level playing field to start with. That is why we have privatization and competition at the two ends and then nationalization in the middle to provide and maintain the infrastructure."

In the United States, the balance between privatization and regulation is not as clearly defined, and many experts in the industry look at rampant deregulation over the past 20 years as the reason why the system is so outdated. Because of governmental deregulation, much of the infrastructure is privately controlled, and few upgrades are being made.

Despite the fact that energy prices have actually gone down over the past 20 years in the United States making production and transmission less expensive, the end result has been more players entering the energy market, which has led to paper-thin profit margins across the industry. Thus, major companies do not have the excess resources to invest in the critical infrastructure (i.e., power lines, grid maintenance, generators) needed to prevent the system from becoming even more archaic than it already is.

In 1965, a similar blackout struck the Northeast, but the 2003 outage was much more far-reaching and dwarfed the previous blackout by comparison. According to Cherry, however, the regulatory response to the 1965 blackout was initially more notable than that which has taken place since last August, and he believes that things will not really begin to improve until the federal government takes a more defined role in managing the national grid.

"There needs to be a conclusion as to which federal agency is responsible for the national grid and what standards will be implemented," says Cherry. "There's a lot of indecision. A lot of people are hesitant to make investments because they don't see any clear path. Until you have [federal government action], there will continue to be a level of uncertainty that will inhibit any major investments towards upgrading the grid."

Greenhill agrees that the problems with the power grid system in the United States are not insurmountable. With a concerted effort by the right governmental regulatory agency and committed investments by the right players in the energy market, the grid can be modernized and future blackouts can be avoided. However, nothing will progress until these hurdles are cleared.

"You can sort all of these problems out," says Greenhill. "Technologically it is not a problem. It is basically all down to money and systems, and that's why there isn't any single technological solution in the United States. But there should be one. The problem is that there's a lot of different self-interest groups in different parts [of] the country, who say that they like to have their own power . . . so it becomes more of a political issue than a technical issue."

Electric Intelligence*

Establishing a Smart Grid Requires Regulatory Reform, Not Subsidies

By Lynne Kiesling
Reason, June 2009

How much thought have you given to your electricity consumption? If it hasn't gone beyond "I flip the switch and the light comes on," you're not alone, which is one of many reasons electricity usage in the United States is inefficient. But that's beginning to change.

The digital communication technology revolution that has created mass productivity gains throughout the economy during the last 20 years is finally creeping into the electricity industry as well, under the broad moniker of the "smart grid." The basic concept of the smart grid is to replace the faceless, impersonal "grid" of networked electricity delivery with a transparent, interactive system that allows users to see and select from pricing choices in near-real time. President Barack Obama included $4.5 billion in smart grid subsidies in the stimulus package Congress enacted in February. But absent substantial regulatory reform, mostly at the state level, such federal spending may simply reinforce a century-old model that is all but obsolete.

Imagine a future in which your home has a system that connects all its appliances, entertainment systems, heating and cooling, laundry, and lighting into one communication network. The network would be accessible through a computer screen or a Web-based portal. Through this interface, your electricity company would communicate real-time information about how much electricity you're consuming, the price you're paying at different times of the day, and whether the juice is coming from renewable or conventional sources.

Using that information, you could change the settings on your various devices in response to prices. If you knew that you could save money by lowering the temperature of your water heater five degrees when the price per kilowatt hour increases from nine to 12 cents, you would be more likely to lower the temperature of your water heater. Furthermore, if you could program your appliances to take care of the price response work for you, you'd be even more likely to do it. And once plug-in electric vehicles become more widespread, you could set up your home network to charge the car during less expensive, off-peak hours, and maybe even sell your excess energy to neighbors when prices are high, since electric vehicles are essentially energy storage devices.

Such systems are becoming increasingly feasible as information technology costs fall. Intelligent devices such as your thermostat, water heater, television, and plug-in vehicle all have digital communication capabilities and can be programmed to respond autonomously to data, including price signals. An in-home system networking these machines could cost as little as $250, with costs going up as sophistication and functionality increase. Having extra intelligence embedded in devices could add, say, $25 to the price of a clothes dryer, to judge from estimates from the production of such dryers for the GridWise Olympic Peninsula testbed project. Home electricity management systems are being developed and will be on the market in the next several years.

The key piece of network infrastructure that's making it easier and cheaper to get data to and from such in-home systems is a "smart meter." A simple digital meter that communicates only with the utility company can cost as little as $50. More sophisticated meters that talk with both the utility and individual customers can cost as much as $300.

The ability to communicate from customer to utility and back will benefit consumers by helping them save money, buy new energy-saving products and services, and reduce their environmental impact. But to make this vision a reality will require regulatory reform, particularly at the state level. Utility regulation has long been based on the now-obsolete idea that electric power generation and delivery is a natural monopoly. Consequently, state regulators have granted most utilities a monopoly over the retail sale of electricity products and services within a geographical area, as well as a monopoly over the construction of distribution wires across public rights of way.

The most crucial regulatory reform would be eliminating the single, fixed retail rate for most consumer electricity consumption. In February, Obama pledged to install 40 million smart meters in homes. But it doesn't matter how smart these meters are if homeowners are going to get charged the same old flat rates. Customers need to know how prices vary over hours, days, and seasons. With that information they can decide how much energy to buy and when. Smart technology makes responding to changes in price as easy as scheduling your DVR to record your favorite TV show.

Some regions already have a rudimentary form of a smart grid in which customers can respond to some prices. One variant called "time of use" lets consumers know in advance what prices will be during certain hours. But the pricing choices offered by real retail competition is rare. In a large number of pilot projects, including the California Statewide Pricing Pilot and the GridWise Olympic Peninsula project, customers typically save 10 to 20 percent on their electricity bills and reduce their peak electricity consumption by 12 to 15 percent. But the bulk of residential customers still pay fixed prices that are calculated by regulators to cover the costs of the utility plus a profit margin.

In such a heavily regulated environment, smart devices offer little if any value to consumers. Technological innovation and regulatory innovation must go hand in hand.

In fact, regulation is the chief reason why the electric power industry is so technologically backward to begin with. Since utilities are regulated monopolies, they don't usually receive strong price signals telling them whether an investment is a good or bad idea. Instead, regulators decide which investments are prudent on consumers' behalf. Regulators can decide that a utility's investment in a groundbreaking new facility or technology is imprudent after it's already been built. Naturally this process makes utility executives (and regulators) conservative; it's just safer to build what has been previously approved.

The long time frame over which utility assets depreciate reinforces that conservatism. Many utilities, for example, still use analog electricity meters that are more than 50 years old, despite the many advantages promised by two-way digital communication. While electric utilities remain mired in regulatory backwaters, many other industries, from travel to retail sales, have been embracing new technologies to enhance communication with customers. The result has been the creation of new products and services, faster economic growth, higher profits for producers, and improved wellbeing for consumers.

Regulated utilities are also accustomed to having control. As owners of power generation plants and electric distribution wires, they have been managing the power network using top-down hierarchical control for almost a century. Not surprisingly, they want to use the new smart grid capabilities for "direct load control"—shutting down your air conditioner from afar during peak hours, in return for which they would offer you a rebate.

Keeping the electricity grid physically balanced in terms of frequency and voltage prevents blackouts (which cost about $160 billion per year) and is the primary performance metric on which utilities are evaluated. While some consumers are OK with having the power company manipulate their air conditioners, others find it downright Orwellian. But the same outcome could be achieved by allowing consumers to receive and respond to real-time price signals themselves, instead of just leaving it all up to the producer. Decentralized coordination is ultimately more efficient and empowering than imposed control.

So what about Obama's smart grid funding? It's likely to be spent as wastefully as your average government subsidy for technology and industry. But the plan might have two benign effects. First, it will induce utilities to make technology investments that might be profitable if not for the perverse incentives created by regulation. Using government spending to achieve this outcome is a way of remedying one policy-induced distortion with another, but in the existing regulatory environment this might be a politically palatable way to overcome the stifling, regulation-induced inertia of both the regulators and the regulated. Second, by reducing the cost of information flow to and from consumers, smart grid subsidies may contribute to the erosion and ultimate disappearance of state-level regulatory barriers to retail competition. Telecommunication deregulation sprang from similarly modest roots.

The obsolete electro-mechanical electric power network, built by and for a monopoly industry, cannot support the kind of growth experienced during the last 20 years in so many other industries. All that stands in the way of vibrant, customer-friendly electricity products and services is an outdated infrastructure run by hesitant monopolies and regulated by bureaucrats with little incentive to improve things. We can do smarter.

LYNNE KIESLING *(lynne@knowledgeproblem.com) is a senior lecturer in the Department of Economics and the Kellogg School of Management at Northwestern University, and a member of the GridWise Architecture Council. She blogs at knowledgeproblem.com.*

Wiring the Revolution[*]

By Thomas K. Grose
ASEE Prism, January 2011

Few would argue against the need to transform America's aging and increasingly decrepit electric-power grid into a more robust Smart Grid, using digital technologies. But ultimately, consumers must buy in to the concept. And for that to happen, emerging technologies will have to be dead-simple to use, generate cost savings, improve efficiency, and "have a cool factor," says Alex Huang, a professor of electrical engineering at North Carolina State University.

That's where the Future Renewable Electric Energy Delivery and Management Systems Center comes in. Directed by Huang, the federally funded center is a seven-university international consortium with 44 industry partners headquartered at NCSU. It seeks to develop and demonstrate technologies so cool that they'll revolutionize the power grid and hasten the day when America can use more renewable fuels to generate electricity. While advancing toward that goal from many directions, the FREEDM Center is focused particularly on three targets: a two-way digital communications backbone for the grid, lightning-fast solid-state transformers, and improved batteries—particularly for plug-in hybrids and electric vehicles. "The technology to do all of this is still not ready," Huang admits. But, he quickly adds, "we are making great progress."

Today's grid is a one-way system. Power companies generate and distribute electricity from large—and in the United States, mainly coal-fired—plants to businesses and households. Because the flow of electrons through the grid needs to remain uninterrupted, the grid's not good at accommodating power from renewable sources, like […] wind and sunshine, because they're intermittent supplies. But using digital technologies and two-way communications to better control, manage, and balance demand and generation will make it easier to bring renewables online. Ultimately, a Smart Grid will also allow users to sell electricity back to utilities from, say, home solar panels or idling electric vehicles, or EVs.

Denver-based Green Energy Corp., a FREEDM Center industry partner that develops communications software for the grid, is heavily involved in designing a two-way communications system. Based on the firm's grid-management software, it's an open-source, cloud-based system that lets utilities upgrade legacy networks with plug-and-play applications from outside companies. "Our technology works with any vendor's technology," says Roxy Podlogar, Green Energy's vice president for product strategy. Two-way communications is also key to relying on renewable fuels, since it allows a grid to automatically and instantaneously switch power sources if one or more start to fade.

A superfast, solid-state, electronically controlled transformer will act as an "energy routing device" between the grid and consumers, the so-called last mile of an intelligent grid, says Huang, who is leading its development. Traditional copper and iron transformers "can change voltage, and that's all." His semiconductor-based transformer will also change frequencies, and connect to both AC and DC devices, including electric vehicles, wind turbines, and solar panels. "It's an enabling technology for a more actively controlled grid," Huang adds. The transformers also will be smaller and lighter than today's, and produce less heat.

Faults, of course, can and will occur in any power system, but they needn't be disruptive. The average U.S. household is without power for around four hours a year; in Japan, the average is a mere seven seconds a year. So the center's researchers are working on a device, a sort of electronic circuit-breaker, that immediately isolates faults and reroutes power. "If you have a fault," Huang says, "the goal is to lose zero customers. The expectation is perfect power."

A Smart Grid system more dependent upon renewables will need to store massive amounts of energy, so battery technology is another area of research. However, for now, the center's focus is mainly on batteries for EVs and plug-in hybrids because of the ongoing push from Washington and the auto industry to make electric powertrains the next big thing. If these cars sell in great numbers, as is hoped, it could affect the grid, especially if too many drivers decide to plug in at once. If, for instance, there are 100 EVs in a parking deck, "that becomes a very complex control problem," Huang says. "Which one do you charge first, which one second, and at what rate? That's a grid control problem."

Center researchers are also using a technique developed at NCSU called electro-spinning that weaves nanofibers into a new composite material for lithium-ion battery anodes, enabling them to store more energy and endure more abuse. And, to be sure, assaults on batteries—and the grid—could occur during fast charging. Ideally, car batteries would be charged slowly, overnight—when demand and rates are low—but obviously there will be times when drivers can't wait that long. Asks Huang: "Can you really charge them in 20 minutes? And how much charge can be put in in 20 minutes and not cause damage?" Those are questions his researchers hope to answer.

The "cool factor" in these novel technologies needs to be demonstrated. So the 20,000-square-foot center has its own 1-megawatt microgrid, which soon will be festooned with all sorts of devices, monitors, control software, and power sources, including EV charging stations and juice from a 40-kilowatt solar array donated by Germany's AEG Power Solutions. Also in the works are plans to link the microgrid to a model "smart building" that IBM is constructing within the center to demonstrate to commercial building owners how energy-management software can cut power usage by 10 percent. "We want customers to come in and say, 'Oh, now I get it,'" explains L. Steven Cole, the IBM program strategy manager who's setting up the site. It also will demonstrate software that lets users switch from in-house solar or microturbine systems to the grid and back. "It's not a simple thing to do," Cole adds. Rogelio Sullivan, the center's assistant director, calls IBM a good fit: "What they're doing parallels our work."

INDUSTRY'S MAJOR ROLE

The FREEDM Center is part of the Obama administration's multibillion-dollar research and development effort to transition the nation into the Smart Grid, or Energy Internet. Funded with an initial five-year, $18.5 million grant from the National Science Foundation, the center comprises NCSU; Arizona State, Florida State, and Florida A&M universities; the Missouri University of Science and Technology; RWTH Aachen University in Germany; and the Swiss Federal Institute of Technology. It's anticipated that the NSF will renew the five-year grant when it expires in 2013. The center is also funded by $10 million that comes from the universities involved and industry fees. So far, it has 44 industry partners. Besides Green Energy, they include such heavyweights as Toyota, Duke Energy, and small start-ups like MegaWatt Solar along with IT giant Cisco. Having businesses play a major role in the center was in the plan from the start. Indeed, Green Energy Chairman Daniel Gregory headed the center's industrial advisory board and helped write the grant proposal.

Huang's confident that the FREEDM Center will demonstrate that the technologies it's developing will work. "A harder question is how much of that will translate to the market," he says. That's a big reason the center was designed to have a strong industrial component. The market challenge is also embedded in another major element of the center: education. NCSU, via the center, offers a bachelor's degree in electrical engineering with a concentration in renewable electric energy systems (REES), and graduate students can earn a certificate in REES. Moreover, next year the school will begin offering a one-year master's degree in electric power systems engineering to professionals looking to update their skills and career prospects.

Center students are expected to be entrepreneurial and have a strong understanding of market forces. Not only are students required to take business courses to learn how to write a business plan and do market analyses, but each of the 20 research projects at the center has an industrial "champion" who works with faculty and students to help keep them commercially oriented. Venture capitalists are brought in to give researchers and students tutorials on launching start-ups.

The so-called soft skills also are stressed. "Students must also have effective communications skills to work with the general public, business partners, and among themselves," explains Leda Lunardi, a professor of electrical engineering and also the center's education director. A useful training ground in these skills is an ambitious outreach operation aimed at middle and high schools that includes sending grad students into classrooms to do experiments and a Young Scholars program that brings students and teachers on campus for five weeks in the summer.

For now, the FREEDM Center has the prevailing political winds at its back. But, of course, that could change. The incoming Congress, for example, is heavily populated with climate-change skeptics. Could that ultimately affect funding for this type of research and sap industrial enthusiasm for Smart Grid technologies? No way, says David Bartlett, an IBM vice president who is overseeing the smart-building project. A Smart Grid offers the power industry too many advantages and cost savings, he says. "Every utility is involved in an upgrade to Smart Grid and digital technologies." Huang agrees. Industry is "genuinely interested" in wanting a major overhaul of the grid, he says. "There is some robustness in this. I think for [industry] this is a long-term business decision." If Huang's right, the FREEDM Center's future as a hotbed of research into cool technologies seems secure.

THOMAS K. GROSE *is* Prism's *chief correspondent, based in the United Kingdom.*

5

From Taps to Toilets:
Waterworks

Courtesy of Robyn Beck/AFP/Getty Images

A worker tightens bolts on giant pipes which will be used to pump water out of the 17th Street Canal in case of rising waters caused by a storm, in New Orleans, Louisiana, on July 1, 2006.

Scenes from an artifical wetland filtration system designed to purify runoff water set in an agricultural area which traditionally uses water from the Colorado River distributed through a series of canals and irrigation channels, in Imperial Valley, California, on August 6, 2009. The Imperial Valley is a desert area which controversially uses three quarters of California's allocated river water for agricultural purposes.

Editor's Introduction

California has perhaps the most precarious water infrastructure in the nation. Two thirds of the state's 37 million people depend on the delta of the San Joaquin and Sacramento Rivers for their water. An elaborate system of pumps, levees, and canals transports the water from the delta on San Francisco Bay hundreds of miles south to the teeming cities of Southern California. Unfortunately, the delta's levees are in disrepair and the delta itself rests on the Hayward Fault, a powerful earthquake zone. Experts fear that even a minor earthquake—or a major storm—is all that it would take to overrun the levees and inundate the delta with salt water from San Francisco Bay, causing a veritable meltdown of the state's water infrastructure and leaving around 25 million people without access to freshwater.

Meanwhile, an ongoing drought and environmental concerns have limited the amount of water that can be pumped from the delta. Farmers have borne the brunt of the water cutbacks and a near permanent state of economic unease has taken hold in the agricultural communities of California's Central Valley, as some of the most fertile land on the planet lies untended.

Potential solutions to California's water woes tend to be costly and complicated. Currently there are plans for a new canal, desalinization and wastewater recycling plants are up and running, and conservation initiatives are in place. Still, for a state that adds hundreds of thousands of new residents each year, whatever measures are implemented may not be enough to quench California's thirst.

Entries in this chapter consider the condition of the U.S. water supply and its accompanying infrastructure—the dams and levees, sewers and reservoirs, that keep the taps running and the toilets flushing. In the first selection, "At the Breaking Point: The Sorry State of the Country's Water Infrastructure and What It Means If We Don't Fix It," Robert Barkin offers his take on the situation, using a water main break in San Antonio, Texas, as the jumping-off point for his examination. He also discusses several water system improvement projects occurring throughout the country.

According to Laura Shenkar, in "Blue Technology Can Quench America's Thirst," the subsequent piece, "the glass—we are happy to report—is half full." While our water management systems are under threat due to underfunding, climate change, and other factors, Shenkar believes that advances in technology will provide the necessary solutions. Jeff Gunderson addresses the problem from an economic perspective in "Not Just a 'Water Problem': Investment in U.S. Infrastructure Systems Is Critical to Keeping U.S. Globally Competitive, Report Says," contending that improving and protecting our water infrastructure is especially vital to our future prosperity.

In "Water at the Crossroads: Twenty-First Century Issues Demand New Approaches to Water Management," Jennifer Grzeskowiak describes various water infrastructure crises throughout the nation and the efforts underway to address them. To Glen Daigger, a technology officer for a water company in Denver, the situation is not as dire as some would believe. "We have a water supply problem only if we think we're constricted by past practices," he observes. "We have a water management problem, not a water problem. The solution is to change how we manage water."

In the final selection, "Add Water Systems to U.S. Infrastructure Challenges, Says *Infrastructure 2010: An Investment Imperative*: New Research Compares Urban Markets in U.S. and Overseas, Urges Changes in Land Development Patterns," a writer for the Urban Land Institute discusses an alarming report on the state of water systems in cities throughout the United States.

At the Breaking Point[*]

The Sorry State of the Country's Water Infrastructure and What It Means If We Don't Fix It

By Robert Barkin
American City & County, January 2009

Residents of a neighborhood on the southeast side of San Antonio awoke Dec. 12, 2008, to a mess outside their homes. A water main had broken, spilling water over lawns and damaging Christmas decorations. The San Antonio Water Systems (SAWS) spokesperson attributed the break to drought and the age of the pipes.

The San Antonio neighborhood was not the only one in the city to experience a water main break last year. According to SAWS, the water district had 800 more breaks in 2008 than in the previous year, an 80 percent jump. And San Antonio's pipe bursts were not isolated. A Dec. 12 Internet search using the term "water mains breaking" found 3,293 items. Granted, the time span was greater than one day, but the sheer volume of news stories on the subject is a good indication of the size of the problem. For example:

- A water main break in Las Vegas flooded the area near the University Medical Center of Southern Nevada, closing a number of streets.
- A water main break in Elyria, Ohio, spewed water onto the Ohio Turnpike, turning one lane into an icy hazard. "This happens all the time in the wintertime," said Elyria's Assistant Safety Service Director Jim Hutchson.

And, in one of the biggest water main breaks in recent history, a 66-inch main broke in Bethesda, Md., making national news just before last Christmas. The break poured 150,000 gallons per minute down a major road, coincidentally named River Road, in a rush that marooned a dozen commuters in their cars until helicopters could lift them to safety. That was only one of 1,709 water main breaks and leaks in Montgomery and Prince George's counties in 2008, the fifth worst year since 1984, according to the Washington Suburban Sanitary Commission (WSSC). WSSC's record for water main breaks and leaks in one year is 2,129, set in 2007. "We're plagued by old pipes," explained John White, spokesman for the local water utility.

The anecdotal evidence is supported by several national studies that have concluded that water infrastructure—for both drinking water and wastewater—is badly in need of an infusion of funding to rebuild aging pipes and plants. Unless the nation undertakes a concerted effort in the next two decades, the studies predict more pipe) breaks and water shortages and a return to the pollution levels last seen before the enactment of the Clean Water Act of 1972, when rivers were catching fire and swimming holes were closed because of excessive bacteria, according to the Alexandria, Va.-based Water Environment Federation (WEF).

Experts from the U.S. Environmental Protection Agency (EPA) and the Reston, Va.-based American Society of Civil Engineers (ASCE) predict that the nation faces a calamity if spending is not increased by as much as $450 billion over the next 20 years to maintain and repair drinking water and wastewater networks. They foresee a future of ruptured pipes and sinkholes, tainted drinking water and sewage-saturated rivers and lakes. "Our nation's water infrastructure needs have grown while federal funding has declined," says Rep. Earl Blumenauer, D-Ore., a long-time advocate of investment in transportation and water systems. "We urgently need a new federal commitment to significantly increase investment in our water infrastructure."

FALLING FURTHER BEHIND

In the most authoritative assessment of the nation's infrastructure needs the EPA published "The Drinking Water and Clean Water Infrastructure Gap Analysis" in 2002 that estimates the gap between historical funding trends and needs from needs from 2000 to 2019 to be as much as $450 billion. For wastewater, the estimates of investment needs and spending used to calculate the gaps cover all of the approximately 16,000 publicly owned treatment works. For drinking water, the analysis covers the approximately 54,000 community water systems and the 21,400 not-for-profit non-community waters systems in the 50 states, I.S. territories and tribal areas.

In its 2002 analysis, the EPA estimated current capital spending at a pace of $13 billion per year for clean water and of $10.4 billion per year for drinking water. The capital payments gap—calculated by subtracting the current spending from the capital payment needs—amounted to as much as $177 billion for clean water and as much as $267 billion for drinking water. Other studies from the Congressional Budget Office, the Water Infrastructure Network and the Denver-based American Water Works Association (AWWA) all have drawn similar conclusions though the actual estimates vary.

"The bottom line is that we will need to spend hundreds of billions of dollars for drinking water and wastewater in the next 20 years," says Tom Curtis, deputy, executive director for government affairs for AWWA, the association representing municipal drinking water utilities. "Drinking water has tremendous needs, and wastewater needs are just as great."

Local governments are spending only 80 percent of what is needed to upgrade their drinking water systems, and the enormous cost of pipe replacement is the leading factor in the shortfall, Curtis says. "That's the lion's share," he says. "People don't realize how expensive these networks are."

The cost involves tearing up streets, laying down new pipes and then repaving. One Midwest city estimated that completely replacing its existing network of pipes would cost $2 billion. "Pipes last a very long time," Curtis says, "but they don't last forever."

Many people think that the replacement issue is confined to older cities in the Northeast, where pipes may be 100 or more years old, Curtis says. But those older east iron pipes have a very long economical lifespan and may last for another 20 years. Newer pipes, made with less expensive materials, may have shorter lifespans. That is why pipes laid 100 years ago may still be in usable shape, while neighborhoods created in the 1950s and 1960s may have bursting pipes today.

As a consequence, Curtis says, over the next 20 years both older and newer cities will be facing replacement costs. "There's a great deal of drinking water infrastructure that will need replacement in a fairly compressed time period," he says. "We will need to be ramping up expenditures on pipe replacement."

For the most part, drinking water systems draw revenues from users through rate assessments and taxing districts. The industry continues to believe that the water users should bear the greatest burden of the cost of infrastructure replacement, Curtis says. However, some water authorities face financial hardship and will need support from federal and state programs.

Without the investment, users will face a "degradation of service," including more main breaks, pipe leaks and diminished water pressure, Curtis says. In addition, economic expansion in many industries relies on the availability of a reliable water supply. Newer issues, such as climate change and drought, bring additional urgency to careful water management. "We are looking out over a long time horizon," he says. "We will be competing for use of every drop."

WASTEWATER FACILITIES ARE JUST AS BAD

The investment needs of wastewater facilities are very similar to the massive upgrade requirements of drinking water facilities, says Tim Williams, WEF's managing director of government affairs. "The infrastructure that was built in the '50s, '60s and '70s is at the end of its design life," he says. "Some are way beyond their useful life and need to be replaced. Cities are just running their systems until they fail."

According to ASCE, the nation could lose much of the gains it has made in improving water quality over the past 30 years unless increased investment is made in the infrastructure in the next 20 years. "The U.S. could wind up with dirtier water than existed prior to the enactment of the Clean Water Act of 1972," ASCE states in its 2005 Report Card for America's Infrastructure.

Williams cites a number of reasons for the need for increased wastewater funding. He notes that there are increasingly stringent regulations that require changes in plants that are decades old, the population base of some cities has diminished so that costs are borne by fewer residents, and the manufacturing base, which has often carried a large share of treatment costs, has withered in some cities. "There needs to be a community to spread the cost around," he says. "It's a real hardship."

Like Curtis, Williams believes that users should pay for the operations and maintenance of the wastewater systems and the substantial majority of the costs of capital upgrades. But, he says that the improvements will move forward faster with federal assistance. In economically depressed cities, the only prospect of completing the necessary renovations would be through grants, because the cities do not have the financial capacity to pay for them.

The 2005 ASCE report card states that older systems are plagued by chronic overflows during rain storms and heavy snowmelt, leading to raw sewage discharges into surface waters. The EPA estimated in August 2004 that the volume of combined sewer overflows discharged nationwide is 850 billion gallons per year. Sanitary sewer overflows caused by blocked or broken pipes release as much as 10 billion gallons of raw sewage yearly, according to EPA.

There are up to $20 billion in capital projects that are already designed and approved and only need finding to begin work, Williams says. "We look at this as capital investment," he says. "It's very expensive capital costs that will have to be incurred anyway. It's better to take care of this sooner rather than later."

MANDATES ADD TO COSTS

One water project that is ready for construction is the regulation-mandated conversion of an open reservoir system in upstate New York to a closed reservoir system, which will be ready for bidding early in 2009, at an estimated cost of $55 million.

The construction program is part of an ongoing capital needs program that is designed to upgrade the Onondaga County Water Authority in the Syracuse area over a 20 years, says Mike Hooker, OCWA's executive director. "This way we can smooth out our cash flow," he says.

OCWA regularly assesses the condition of its water mains throughout its network. Much of the infrastructure is 100 years old, while other parts are much newer. Ironically, the cast iron pipe is sometimes in better condition than the newer pipe, because of the better quality of the older pipe or the acidic nature of the soils in the newer areas. "We invest $7 million a year on average," Hooker says. "Everything we make from revenues we put right back into the system."

Hooker, who chairs AWWA's water utility council, is very aware of the needs of the water system nationwide. "The cheapest way to protect our water supply is not to let it get contaminated in the first place," he says.

Both AWWA and WEF advocate an expansion of the state revolving loan programs, which receive federal funding for low-cost loans to the drinking water and wastewater providers. Over the life of the loans, the savings can be millions of dollars, but the limited availability of the revolving line of credit means few projects can move forward, according to the organizations.

ASCE advocates an annual appropriation of $1.5 billion from the federal general fund to the state revolving loan programs. ASCE also supports a significant increase in research funding to develop more cost-effective means of improving the nation's water system. "Next to air, water is the most important resource on earth," Hooker says. "There's no substitute for it."

SAN FRANCISCO OVERHAULS AGING,
EARTHQUAKE-VULNERABLE SYSTEM

Project: **Water system improvements**
Jurisdiction: **San Francisco Regional Water System**
Agency: **San Francisco Public Utility Commission**
Vendor: **Sunnyvale, Calif.-based Trimble, others**
Scheduled for Completion: **2014**
Estimated cost: **$4.3 billion**

Built in the early 20th century, the San Francisco water system is near three major earthquake faults. Recognizing the risk of earthquake damage, the San Francisco Public Utilities Commission launched a $4.3 billion project to repair, replace and seismically upgrade the system's deteriorating pipelines, tunnels, reservoirs, pump stations, storage tanks and dams.

Six years after voters approved a bond measure to fund the initiative, the Water System Improvement Program (WSIP) has identified 85 projects, with completion targeted by the end of 2014. As of Oct. 4, 2008, 23 WSIP projects had been completed. With the recent approval of environmental reviews, another 17 projects totaling $1.8 billion now can proceed.

The San Francisco Regional Water System serves more than 2.5 million customers in four counties. Some districts have begun raising rates, which are anticipated to remain at or below the state median after the project is completed.

A key focus of the project is to ensure water availability after a major earthquake, which, from any of the faults, would likely cut off most customers from water service for at least 30 days. One of the primary goals of the WSIP is to deliver water to 70 percent of customers within 24 hours of a major earthquake.

REMEDIATION REVITALIZES TAMPA BAY DESALINATION PLANT

Project: **Desalination plant**
Agency: **Tampa Bay Water**
Contractor: **American Water—Pridesa (a partnership between Voorhees, N.J.-based American Water and Madrid, Spain-based Acciona Agua)**
Date completed: **Spring 2007**
Cost: **$110 million for initial cost and $48 million for remediation**

After a history of turbulence, the Tampa Bay Seawater Desalination Plant is up and running as the largest desalination plant in North America, providing 10 percent of Tampa Bay Water's supply. The new water supply is a welcome relief for the fast-growing area, which previously relied on groundwater for its drinking water.

Because of design and financial problems with the original contractor, the facility never fully went online. Tampa Bay Water shut down the plant in June 2005 and hired a new contractor to correct the processes.

With an expected lifespan of 30 to 50 years, the facility supplies 25 million gallons per day (mgd) and can expand to 35 mgd. The average unit cost of the water produced is less than a penny per gallon. The water is distributed to Tampa Bay Water's six member governments, which supply the water for more than 2.5 million residents in the region.

The plant removes salt from seawater through reverse osmosis, which uses high pressure to force pretreated water through semi-permeable membranes that trap salt and other minerals. The treated water is combined with water from other sources at Tampa Bay Water's facilities. The salty water residue created by the process is diluted with water used at a local power plant and re-enters Tampa Bay at near-normal salinity.

REGULATIONS PROMPT STORAGE FACILITY CONSTRUCTION

Project: **Conversion to closed reservoirs**
Agency: **Onondaga County, N.Y., Water Authority (OCWA)**
Engineer: **White Plains, N.Y.-based Malcolm Prime**
Anticipated completion: **Summer 2010**
Estimated cost: **$55 million (Bidding expected in early 2009)**

OCWA is replacing its open reservoirs with closed reservoirs, which will meet. EPA standards to protect against viruses and parasites that have periodically infiltrated systems in other parts of the country. A particular problem is Cryptosporidium, a protozoan parasite that affected 400,000 people in Milwaukee several years ago.

New regulations require open reservoirs to convert to closed systems to keep out biological waste from birds and animals. At the same time, OCWA will use the $55 million project to redistribute water resources closer to the locations of great usage.

According to the plan, a 50 million gallon open reservoir will be drained and replaced a 20 million gallon closed reservoir on the same site. On the other side of the district, where population has grown, a 30 million gallon tank and a 20 million gallon tank will replace an open 30 million gallon reservoir. The better allocation of water storage will keep the water fresher, says Mike Hooker, OCWA's executive director.

The project, which is designed and ready for bidding, will take an estimated 18 months, with completion anticipated for summer 2010. Using the state revolving fund for financing, the project will save $2 million over the life of the project.

Blue Technology Can Quench America's Thirst*

By Laura Shenkar
USA Today (Periodical), September 2009

Is the water glass half empty or half full? It is half empty because of our future water shortages, increase in pollution, and health risks at our beaches all occurring within an economic climate that does not recognize water as a priority.

One of the great challenges that the U.S. faces over the next decade will be ensuring an ample supply of quality water. Even under normal weather and water conditions, water managers in 36 states anticipate water shortages within the next five years. Water usage patterns become a factor here. New Englanders use less than 20 gallons per day per person, while Southwesterners consume about 200 gallons per person per day. Population growth in the driest places has continued to expand, depleting water supplies rapidly while creating even more runoff pollutants.

At the same time, the pipes that deliver and remove water from our homes and businesses are collapsing as a result of decades of neglect. The U.S.'s drinking water piping network extends more than 700.000 miles and much of the infrastructure is at least 100 years old. With over 1,200,000.000 miles of sewage piping in the U.S., storm water and sewage collection systems fail daily, stopping business and creating significant health hazards.

Finally, climate change brings in a new element of uncertainty for water—potentially increasing the number of severe storms, floods, droughts, and heat waves, thus making our existing centralized water treatment and delivery for the steady supply and quality of water that are essential for human health and for economic activity even more problematic.

When rain falls on paved or hard surfaces, instead of natural environments, it carries oils, wastes, and contaminants into rivers, streams, and beaches—a process known as runoff. Until the 1970s, the disposal of toxic water was considered harmless. It was thought that any polluted discharges would be diluted by the significant water volumes in the oceans. However, through rapid development, our communities have become denser and storm water pollution more dramatic. The General Accounting Office estimates that 46,000.000 gallons of oil are spilled, dumped, or run off into the nation's rivers, lakes, streams, and oceans annually.

Beach closings and advisories are at a record high for U.S. coastal waters. When it rains heavily, untreated sewage combines with contaminated storm water to flow into waterways from overflowing sewers, storm water pipes, and treatment plant bypasses, carrying human and animal wastes, trash, and other bacteria-laden discharges into the water. The most common health impact of this contamination is severe stomach upset, but swimmers also can get earaches, pinkeye, respiratory ailments, and even more serious illnesses like meningitis and hepatitis. Small children, the elderly, pregnant women, cancer patients, and those with weakened immune systems are most at risk for illness caused by contaminated beach water. According to the Centers for Disease Control and Prevention, children under the age of nine had more reports of diarrhea and vomiting from exposure to waterborne parasites than any other age group.

When the monitoring of water at swimming beaches indicates that levels of certain indicator bacteria exceed safe standards, state or local agencies notify the public of potential health risks in accordance with the Federal Beaches Environmental Assessment and Coastal Health (BEACH) Act of 2000. These notification actions usually are in the form of a beach advisory (warning people of possible risks of swimming) or actual closing. Because of a lack of real-time monitoring technology, many states close beaches preemptively in anticipation of heavy rainfall or other seasonal factors.

Health care outlays for beach pollution illnesses are estimated to range from $21.000,000 in direct costs to $414,000,000 annually in overall costs, depending on the method of reporting used. For instance, bacterial pollution sickens as many as 1,500,000 swimmers and surfers annually at many Southern California area beaches, spanning 100 miles of shore in Los Angeles and Orange counties. Nearly 80,000,000 visitors recreate at those beaches each year. An environmental science and technology study by UCLA and Stanford University found that between 627,800 and 1,479,200 "excess" cases (beyond the baseline number that normally would be expected) of gastrointestinal illness occur at these beaches every year.

Beaches, rivers, and lakes are the number-one vacation destination for Americans. Each year, we take more than 1,800,000,000 trips to fish, swim, boat, or just relax—an average of approximately six trips per person per year. About one-fourth of the population goes swimming in our waterways annually. Economic activity associated with the ocean contributes around $125,000,000,000 to the U.S. economy on an annual basis.

Coastal states generate about 85% of all tourism dollars. For the towns that rely on vacation dollars for their economy, polluted water places revenues and jobs at risk. Polluted waterways also can reduce the property value of nearby houses and lands. An American Housing Survey found that, all else being equal, the price of a home within 300 feet of a body of water increases by up to 28%. Clean water matters. Higher property values are associated with proximity to beaches and open water, and the fact that people are willing to pay more to be closer to these attractive environmental features.

True, too, is that pressure is increasing on coastal and water resources. According to a report by the National Oceanic and Atmospheric Administration, the number of homeowners along the 673 coastal counties of the U.S. grew 28% from 1980 to 2004. At present, 53% of the country's population lives on 17% of the total land area of the lower 48 states (discounting Alaska and Hawaii). Another 12,000,000 Americans are expected to move to the coasts by 2015.

Advanced technologies give us new solutions to water pollution, often making old practices safe and efficient. There are various ways to prevent much of the pollution and economic damage that storm runoff causes. For instance, rain can be caught as it falls and employed for landscaping and other nondrinking uses. When rainwater is utilized on-site to irrigate landscapes or fill toilets, water that comes through municipal pipes can be used more for drinking and washing. Bringing back the old concept of rainwater harvesting is a great opportunity, but eliminating the disease that plagued the old system is the real challenge.

Over the last few decades, water professionals have focused on developing large water treatment plants with extensive piping networks. They centralized solutions for clean water, and succeeded in improving the quality and dependability of water supplies dramatically. This centralized approach has enabled trained personnel to apply the best methods and most efficient practices with great precision for large populations. However, rainwater harvesting works best on-location outside of our highly controlled centralized water facilities. Consequently, the solutions needed to control storm water pollution and address water scarcity require more than just innovative technologies. The methodology used by water professionals must be designed into new products, in the same way that cars are designed to guide their drivers, such that anyone with just a little bit of training can operate them.

Advanced water technology—quickly becoming known as "blue technology"—is an important and promising industry that should enable us to borrow knowledge from consumer products, industrial design, and advanced natural sciences to provide leading-edge solutions to the widespread water dilemma.

Four main technology product opportunities address the challenges in decentralizing water infrastructure: real-time monitoring at the point of use; on-site wastewater reclamation; advanced membranes to enable cost-effective means for saving energy and water, and more efficient cooling systems. These technologies have the capacity to protect the health and wellbeing of our nation as well as promote economic growth. Innovative products are needed to shore up our deteriorating centralized infrastructure and provide new and sustainable decentralized approaches that are the key to protecting our waterways and beaches. The costs to our communities and environment of not addressing water scarcity and runoff pollution are too great to ignore. These blue technologies need government support—through new laws and funding.

Numerous studies demonstrate that every dollar invested in public water and sewer infrastructure and services yields approximately $8.97 for the national economy. A variety of these blue technologies already are on the market. So, when it comes to the promise of advance water technology, the glass—we are happy to report—is half full.

LAURA SHENKAR *is principal at The Artemis Project, a consulting practice that combines business development for pioneering products with projects that apply water conservation approaches to large corporations.*

Not Just a 'Water Problem'*

Investment in U.S. Infrastructure Systems Is Critical to Keeping U.S. Globally Competitive, Report Says

By Jeff Gunderson
Water Environment & Technology, September 2010

In a recent report, the Urban Land Institute (Washington, D.C.) and Ernst & Young (London) addressed the state of U.S. infrastructure, examining looming challenges related to water delivery, energy, transportation, and measuring America's infrastructure investment compared to competing countries around the world. *Infrastructure 2010: An Investment Imperative*, the fourth in a series of infrastructure studies produced by the partnership, explores global infrastructure trends and highlights the need for the U.S. to begin viewing infrastructure as an investment.

Could crumbling infrastructure be detrimental to the U.S. economy? Indeed, the report declares that the U.S. is falling behind global competitors—countries such as China, as well as others throughout Asia and Europe, that recognize the infrastructure investment imperative. America must now act and take a similar approach, the report says, by building the infrastructure investments that are needed to ensure future economic prosperity.

Communities in nearly every region across the U.S. face challenges associated with water, the report says, stemming from problems related to aging infrastructure, contamination threats, shortfalls in conservation, and insufficient supplies for meeting future demand, while globally, countries cope with problems related to drought, inadequate systems, water quality, and pollution.

FUNDING GAPS

Insufficient investment in transportation, water, and other systems during the past 30 years has left the U.S. in imminent need of updating and rebuilding its outdated infrastructure systems, the report says, along with making the necessary upgrades and expansions to support a rapidly increasing population. However, despite this necessity, infrastructure spending is still a low political priority, with other national agendas, such as health care and defense, taking precedence.

The U.S. is falling more and more behind on infrastructure spending as a portion of gross domestic product (GDP), according to Michael Lucki, global leader of infrastructure and construction at Ernst & Young and a primary editor of the report. "In the 1950s and early 1960s, America invested a much larger percentage of its GDP toward infrastructure, and that made the U.S. globally competitive," he said. "But since that era, America's commitment toward infrastructure spending has dropped off significantly. As a comparison, the U.S. now spends only 2% to 3% of its total GDP on infrastructure, while China dedicates upwards of 7% to 9%."

America's infrastructure spending has not kept pace with its growth, Lucki added. "For example, since 1960, less than 10% of new roads were added to the U.S. transportation system, but the number of miles driven on roads has more than doubled over the same time period," he said.

According to the report, while the U.S. struggles to gain traction in planning and building the critical infrastructure investments that are necessary to support economic growth, other nations in Asia and the European Union, despite coping with recessionary-like conditions, can front-load stimulus spending on national and regional infrastructure initiatives already under way, expanding high-speed rail networks or expediting energy and water projects. The report warns that further delay in U.S. infrastructure spending means losing more ground to these countries.

"Of the $787 billion comprising the American Recovery and Reinvestment Act, only $132 billion—or 17% of the total—was allocated toward infrastructure," Lucki said. "It is imperative that we educate people about the dire situation of our infrastructure systems and change our pricing approach for the use of it. Until we do that, infrastructure will remain less of a national concern."

But at the same time, America's infrastructure dilemma presents an economic opportunity. "Infrastructure spending has a multilayer effect," Lucki said. "It helps create jobs, building tax revenues and contributing to a more viable economy.

The U.S. must find new ways to finance repairs and upgrades, according to a recent infrastructure report.

Different studies have shown that every dollar dedicated toward infrastructure can generate approximately $1.75 of additional GDP growth. But without the needed funding and investments, the U.S. infrastructure system could eventually become akin to that of a second—or third-world country."

NEW MINDSET NEEDED

The report examines water issues and concerns affecting 14 metropolitan areas across the nation and concludes that no major metropolitan region in the U.S. is insulated from water-related problems and costs.

"Along with transportation, we have gotten used to subsidized water delivery and wastewater treatment," Lucki said. "But at some point, Americans are going to have to realize the true costs associated with keeping these systems running."

The American Society of Civil Engineers (Reston, Va.) 2009 report card on infrastructure gave letter grades of D- for both drinking water and wastewater, Lucki added, while also projecting a 5-year funding shortfall of $108.6 billion for drinking water and wastewater infrastructure combined.

Underinvestment and a lack of maintenance, according to the report, have left many of the nation's 16,000 wastewater treatment plants deteriorating and unprepared to handle projected demands in the future. "The U.S. water and wastewater infrastructure system is highly underfunded, and moving ahead, it is crucial that the U.S. finds a different way to finance infrastructure repair and upgrade projects as compared to the way it was done over the last 50 years," Lucki said.

Another key aspect is water conservation. "It's going to be highly critical that people change their habits toward water," Lucki said. "The way to change that perception is to raise the price of water, incorporate variable pricing strategies, or bring smart metering into the homes so that people will know the true costs associated with their water usage. By monitoring and measuring, it can be possible to solidify a reaction and, in effect, impact people's behavior. Moving ahead, we need to establish water usage that is aligned with growth."

EASING THE BURDEN

With deficits at the federal level exceeding $1 trillion and lower tax revenues being generated from lower employment, finding the capital necessary for funding infrastructure investments is difficult, Lucki said.

However, the report highlights the advantages of public-private partnerships (PPPs), which can attract the needed investor capital for financing infrastructure while also helping to bring projects on-line quicker, managing systems more efficiently, and employing innovative operating technologies faster than government agencies.

"PPPs have been used in Europe, Australia, and Canada for the past 10 to 15 years as a vehicle for allowing infrastructure to be built and owned, whether it's wastewater treatment, high-speed rail, water delivery, or mass transit," Lucki said. "In those countries, PPPs have been known to provide as much as 15% to 20% of infrastructure financing annually."

At the same time, the U.S. Environmental Protection Agency (EPA) released a new policy earlier this year that directs $3.3 billion in federal funding to programs under the Drinking Water State Revolving Fund and the Clean Water State Revolving Fund to be used toward water and wastewater infrastructure. EPA emphasized that to use this funding, states must implement "smart-growth" principles, and prioritize projects that work to repair and upgrade infrastructure located in city urban centers over water and wastewater extensions in suburban areas that would serve new growth.

SPREADING THE WORD

Programs such as the Water Environment Federation (WEF; Alexandria, Va.) Water Is Life, and Infrastructure Makes It Happen initiative (www.wef.org/wil. aspx) are aimed at helping utilities educate stakeholders about infrastructure issues. Water Is Life targets ratepayers, local leaders, and the media, teaching the value of water infrastructure and the importance of investing in its long-term stability.

Linda Kelly, WEF managing director of Communications, said the program provides outreach materials for utilities and communities for building awareness and support of infrastructure investments. "Long-term education and support from the public is essential to creating sustainable water infrastructure," she said. "We now have more than 400 communities actively engaged in infrastructure education efforts throughout the U.S."

Kelly said these communities are dealing with aging infrastructure in the context of climate change and increased water quality regulation, and, as such, it is important to generate support for upgrading and replacing infrastructure.

"We believe that increasing awareness of the benefits of clean, safe water, plus increasing understanding that infrastructure systems provide it, equals a better chance the public will support the necessary rate increases," Kelly said.

Water at the Crossroads[*]

Twenty-First Century Issues Demand New Approaches to Water Management

By Jennifer Grzeskowiak
American City & County, October 2009

Today's water systems have progressed far beyond their original missions of preventing disease and providing basic necessities, but, for a host of reasons, they are facing new challenges that are forcing some water officials to move past 20th century approaches to water management. This historic shift in the way water use is viewed and managed takes more than the resources of the average water department. It requires a change in the way communities think about Water use—from government leaders who set policy to consumers, including businesses and residents.

This century problems are different than those that affected previous generations. Water shortages are threatening as the population increases and shifts toward urban areas. Environmental issues, such as reducing energy use and minimizing contaminants that enter rivers and oceans, affect entire regions, not just individual cities. Aging water infrastructure just exacerbates the problem, as does residents' resistance to rate increases in their water bills. "It's the perfect storm," says Andrew Kricun, deputy executive director for Camden County, N.J., Municipal Utilities Authority. "And then, you also have the impending retirement of the Baby Boomers, which affects the management [staff's] institutional knowledge."

Technology has helped address many of those issues, but simply implementing new technology is not enough to resolve long-term problems resulting from past decisions. To do that, many water officials are reevaluating how water is acquired, distributed, used and reused, and how those steps can't be altered to create a healthier, more efficient cycle to meet their areas' complex water needs.

LIGHTENING THE LOAD

Take Philadelphia as an example. One of the city's biggest challenges is managing stormwater, which traditionally has resulted in combined sewer overflows (CSOs) following heavy rain. "The problem is that, while it was OK 100 years ago and even 50 years ago to overflow sewage into rivers and streams, that's no longer acceptable," says Howard Neukrug, director of the Office of Watersheds for the Philadelphia Water Department.

To prevent CSOs, the city is looking at its stormwater management system differently. Rather than relying solely on large, costly infrastructure like underground tanks to retain excess stormwater and slowly release it, Philadelphia's goal is to reconnect the natural link between land and water so that "green" infrastructure becomes the city's preferred stormwater management systems, Neukrug says.

To reach that goal, the city is increasing pervious surfaces by planting street trees, increasing green-and-open space, installing permeable pavement and building green roads. The city has launched a comprehensive "Green City, Clean Waters" program that addresses the importance of stormwater management using green methods across all sectors of the city, including streets, schools, parking lots, industry, homes and more. Over the next 20 years, the city plans to invest $1.6 billion to create a green stormwater infrastructure. Philadelphia's overarching Greenworks initiative also supports the water department's goal of creating 3,200 acres of additional pervious surfaces by 2015.

The goals build on current actions. For example, in 2006 the Philadelphia Water Department began a rain barrel distribution program so residents could help capture stormwater on their properties before it enters the system. To date, the city has given out 1,544 rain barrels to residents that attended mandatory 90-minute workshops on how to use the barrels and their benefits for the environment, resulting in the capture of an estimated 5.3 million gallons of water per year. Philadelphia also has installed numerous tree trenches that each manage 1,436 square feet of impervious area, and the city has used pervious materials when streets and sidewalks are replaced through other projects, like street repairs.

"We're cataloging all the land in Philadelphia that is impervious and looking at how we can change that to value rain water, reuse it, infiltrate it and evapotranspire it through greenery anti trees," Neukrug says. "We should have looked at this 150 years ago. But then, it seemed easier to move water away from us."

DECENTRALIZING THE TREATMENT

Nearly 40 years ago, the Clean Water Act of 1972 encouraged cities to build large central wastewater treatment facilities funded by the federal government, with the goal of improving water quality in streams and rivers. Although central facilities helped improve surface water quality, "it took small plants offline and replaced them with large treatment plants," says Paul Brown, executive vice president for Cambridge, Mass.-based CDM.

Now, many water officials are changing tactics and establishing smaller operations throughout the utility area that reduce the distance water travels through pipes. Decentralization is helping save energy lay reducing the demands to transport water to and from a central treatment facility, which accounts for a large portion of a utility's costs. "The less you have to pump and treat, the better off you are from the standpoint of an energy bill," says Glen Daigger, chief technology officer for Denver-based CH2M HILL.

Historically, decentralization has been used for disposal systems, such as septic tanks, Daigger says. Yet, those methods dot not permit wastewater recycling, which has become more available recently because of technological developments. "You can take water that has been used and clean it to almost any quality," he says. "The decentralized technology available allows its to do that efficiently and also cost effectively on a smaller scale."

Membrane bioreactors are one example of that technology. "Membranes are revolutionizing the quality of water that can be provided," says Todd Danielson, manager of community systems for Loudoun Water in Loudoun County. Va. "There are buildings in Manhattan putting membrane bioreactors in basement to provide water for toilets."

Loudoun Water took advantage of membrane bioreactor technology when it built a satellite wastewater treatment facility to supplement the treatment provided by the District of Columbia Water and Sewer Authority (DC WASA). Loudoun Water had reached its treatment volume limit with DC WASA, which could not expand, so it opened its new Broad Run Water Reclamation Facility in May 2008.

The membrane bioreactor is coupled with an activated carbon filter and ultraviolet disinfection process to meet stringent quality requirements. More than just treating the water, Loudoun Water also is looking at ways to reduce the energy used in the process. For example, it captures methane gas produced from the digestion of solids and uses it to run a small boiler. As a reclaimed water use system develops around the facility, Danielson says the utility also can start looking at ways to use the geothermal heat created. (For more information on the facility, see sidebar, below.)

Some businesses and subdivisions already have expressed an interest in using the facility's reclaimed water. "Within a two-mile radius, we can easily construct a reclaimed water network and provide water to these customers," Danielson says. The utility's goal is to use 30 percent of treated effluent for customer's non-potable water needs by 2015.

The utility also recently received $1.78 million in stimulus finds to facilitate reclaimed water use. Loudoun Water will use the Funds to construct two sections of purple pipe that will transport reclaimed water to a private office building and to the National Rural Utilities Cooperative Finance Corporation. Both will use the water for irrigation and cooling towers, among other uses. Danielson hopes to tie other customers, like a nearby golf course, into those lines eventually. "If we're now spending less money on conveyance, we can spend more on reclamation," Daigger says. "We need reclamation to be coupled with increased water efficiency."

CENTRALIZING THE COMMUNITY

As Loudoun Water demonstrates, a 21st century water management model will not be effective without participation from businesses, residents and community groups that are willing to participate in new water initiatives. In a growing number of communities, groups that support efforts to create sustainable cities are easy to find. "In this severe economic downturn, the excitement for the sustainability movement hasn't died down," Neukrug says. "It's remarkable and says that we're maybe really there, and it will be different than in the 70s and 80s."

In Philadelphia, homeowners play a major part in the city's stormwater management plan. In fact, the city had to put a cap on its Model Neighborhoods pilot program, which involves residents in stormwater and watershed improvement projects, because of its popularity.

Involvement from residents also reconnects them more directly to the water cycle—a connection that is lost in most current water treatment networks. "The systems that we have in place are almost designed in a way for the customer to completely take them for granted," Brown says. "They don't know where water comes from, and they don't know where it goes. They just pay the bill."

To complement resident's and businesses' voluntary efforts, cities can adopt building codes that require low-water-use plumbing fixtures, and on-site wastewater and stormwater capture and treatment. In 2006, Philadelphia changed its stormwater regulations to require new developers to capture the first inch of rain that falls. "If that's all that happened, over 60 to 70 years, we would see 40 to 50 percent of private land be stormwater-managed," Neukrug says. "That may not be acceptable as the only thing, hut it becomes a nice part with no direct cost taken up by the ratepayer." Starting in 2010, the city also will start charging non-residential customers for stormwater management based on the amount of impervious surface on their properties.

MANY OPPORTUNITIES FOR CHANGE

With such a confluence of factors contributing to the nation's changing water management needs, numerous opportunities exist for utilities to operate differently. The population may be increasing, but residents are becoming more environmentally conscious and active. Infrastructure may be aging, but the technology to take the burden off of central systems continues to improve. The water supply remains limited, but reusing a portion of it is becoming easier.

"These are challenges that have solutions," Daigger says. "We have a water supply problem only if we think we're constricted by past practices. We have a water management problem, not a water problem. The solution is to change how we manage water."

CASE STUDIES

Loudoun Water
Loudoun County, Va.
Broad Run Water Reclamation Facility

In May 2008, Loudoun Water opened the 11-million-gallon-per-day Broad Run Water Reclamation Facility (WRF) to expand the community's wastewater treatment capacity. The facility is now treating all wastewater for the eastern part of the county and aims to increase reclaimed water use in the area.

The facility uses membrane bioreactor technology to remove biological nutrients. The water then goes through an activated carbon filter and ultraviolet disinfection process. Reclaimed water for use in irrigation or groundwater recharge is drawn off before the activated carbon. The treated water is discharged into Broad Run, which connects to the Potomac River less than five miles downstream. Last year, the facility treated and released 760 million gallons of reclaimed water without a perm it violation.

As part of its sustainability efforts, Loudoun Water has been working to increase reclaimed water use in its service area. "We're trying to find ways to treat water and reuse it so that we have efficient systems for small subdivisions," says Todd Danielson, manager of community systems for Loudoun Water.

Camden County Municipal Utilities Authority
Camden County, N.J.
Delaware No. 1 Water Pollution Control Facility upgrades based on environmental management system

The Camden County Municipal Utilities Authority (CCMUA) treats 58 million gallons of wastewater per day and operates the Delaware No. 1 Water Pollution Control Facility, New Jersey's second-largest wastewater treatment plant. Nearly 10 years ago, CCMUA implemented an environmental management system (EMS) at the plant, which emphasizes reducing the environmental impact while operating more efficiently. Last year, the utility made major upgrades to the facility, helping to achieve the ongoing EMS-related goals of optimizing plant performance, improving odor control and minimizing operating costs.

CCMUA's approach has been to rebuild the plant—constructed in the mid-1980s—project by project using low-interest revolving fund loans, with the intention of reducing operating costs by increasing efficiency. In that way, the utility has nearly completed the major upgrades the plant will need for the next 20 years. Meanwhile, it has kept rates steady for the past 14years.

Last year, CCMUA upgraded the plant's sedimentation tanks, Installed new aerator blades to boost oxygenation, and added a new gravity belt thickener and belt filter presses, reducing sludge disposal costs and improving effluent quality. At $10 million, paid for with a 1 percent interest loan from the state, the sedimentation tank upgrades were the most substantial. "By increasing reliability, we're improving effluent quality," says Andrew Kricun, CCMUA's deputy executive director.

Next year, CCMUA will begin work on the final three projects needed to finish rebuilding the facility. One is a $27 million sludge drying facility that will turn the sludge into a bi-product that can be used as a substitute for coal. It will reduce operating costs by $2 million, and the debt service will be lower than the annual cost savings.

"The EMS helps optimize internal efficiency," says Andrew Kricun, CCMUA's deputy executive director. "You take the private sector model of efficiency and graft that on to the public sector."

"Underperforming equipment is usually more expensive to maintain," Kricun says.

The utility replaced conveying arms with new equipment and swapped wood and steel parts with plastic ones. CCMUA also installed a system that will prevent a catastrophic failure if the chain overruns.

Philadelphia Water Department
Model Neighborhood program

Philadelphia launched its Model Neighborhood program this year, which aims to involve residents in stormwater management and reduce combined sewer overflows. "We're looking at ways to keep water out of the sewers in the first place," says Howard Neukrug, director of the Office of Watersheds for the Philadelphia Water Department. "We are looking at every acre of land and square foot in Philadelphia and seeing how we manage that."

At the residential level, other efforts, like rain barrel use, already have proven successful. Since 2006, the city has distributed 1,544 rain barrels, which prevent an estimated 5.3 million gallons of water per year from entering the city's combined sewer system. To receive a barrel, residents must attend a 90-minute workshop.

Under the Model Neighborhood program, the Philadelphia Water Department is analyzing the specific needs of the blocks involved and then making the improvements necessary to turn them into showcases of effective stormwater management. Changes to the streets include the addition of street tree trenches, bump out and curb extensions, and porous pavement. The department also will recommend measures homeowners can take on their properties to help reduce stormwater.

To become a Model Neighborhood, at least 75 percent of residents on the block must sign a petition. Already, 14 neighborhoods have joined the program, and the department recently had to stop accepting applications because of the large response. Neukrug says the enthusiasm has been encouraging, especially because one goal of the program is to educate residents about the need for greener infrastructure and what forms it will take.

Model Neighborhoods also emphasizes the importance of a holistic approach to stormwater management that involves the entire community, as well as other city departments. The water department has worked with Fairmount Park, PennFuture and Pennsylvania Horticultural Society, among others, to implement the program. "The final step isn't just about water," Neukrug says. "It's about urban life and the quality of life in an urban setting."

Add Water Systems to U.S. Infrastructure Challenges, Says
Infrastructure 2010: An Investment Imperative[*]

New Research Compares Urban Markets in U.S. and Overseas, Urges
Changes in Land Development Patterns

Urban Land Institute (ULI), April 13, 2010

More and more urban areas throughout the United States—in both dry and rainy locales—are facing growing pressures on their water infrastructure systems, necessitating both greater investments for overhaul and a change in development patterns that are more conducive to conservation, according to Infrastructure 2010: An Investment Imperative, a new publication released today by the Urban Land Institute and Ernst & Young.

Citing "water profligacy as an American way of life," the report cautions: "Most water districts do not charge ratepayers full outlays for constructing and maintaining systems . . . As a result, businesses and households tend to use water inefficiently and don't conserve, even though per-capita water demand could outstrip future availability in some parts of the country . . . We are starting to see the limits of where people can go (to live)." The supply/demand conundrum, it notes, stretches from arid California, Colorado and Arizona to humid Georgia and Florida. The report shows that the U.S. has the highest "water footprint" in the world, using nearly 656,000 gallons per capita annually, greatly outstripping far more populous China, which uses less than 186,000 gallons per capita annually.

The integration of more concentrated land development into water management can reduce runoff and combat waste, states the report. One example: the runoff from eight homes on eight acres totals 149,600 cubic feet per year, while the runoff from eight homes on one acre totals 39,600 cubic feet per year, with the denser development saving both water and land. "Changing growth patterns in response to dwindling resources will not come easy to a nation that is not accustomed to conserving water or land," said ULI Executive Vice President Maureen McAvey. "But it's clear that regional and local problems with both water quantity and quality will continue without a broad-based cutback in public water consumption and a change in how and where we build. Water infrastructure must be viewed through the lens of sustainable growth."

"Over the past several years that we have been co-producing this report, perhaps the most troubling conclusion overall is that the world is moving ahead in rebuilding and expanding its infrastructure without the United States. Bottom line, the U.S. is seriously threatening not only its quality of life now and for the future but also its very basic ability to compete economically with the rest of the world," said Howard Roth, Global Real Estate Leader, Ernst and Young.

"Perhaps the priority in the U.S. should be a major jobs-producing investment, aimed at rebuilding national water, transportation, and other life support systems? What ever the solution, this latest real estate report clearly shows that the rest of the world is gaining ground, and that China and the EU among others are already well on the road to recreating themselves as leaders for the new world order. The U.S. has to stop treading water and start treating water," Roth added.

"It's no secret that America's infrastructure is in desperate need of repair, but the real problem is in what you can't see," said Michael Lucki, Global Leader of Infrastructure and Construction for Ernst & Young. He added, "No other infrastructure category presents greater challenges than water. The decisions we make now will impact future generations for years to come."

The report points out that according to the World Bank, 80 countries have water shortages that threaten health and economies, and 40% of the world has no access to clean water or sanitation. Water supply cannot keep pace with demand as populations increase—creating an acute problem in America and worldwide.

Infrastructure 2010 is the fourth of an annual overview series that analyzes the infrastructure needs and compares the infrastructure policies of the United States with other countries. Previous editions focused primarily on transportation systems, consistently finding that the U.S. continues to lag behind Asia and Europe in investments in transit systems, making its urban areas less competitive globally. This year, in addition to a transportation update, the report includes a look at water infrastructure—accessibility and availability, treatment and delivery—and highlights water issues in 14 U.S. cities as illustrative of the problems looming throughout much of urban America. The cities: Atlanta, Boston, Chicago, Denver, Houston, Los Angeles, Miami, Minneapolis/St. Paul, New York City, Philadelphia, Phoenix, San Francisco, Seattle and Washington, D.C. Together, these cities and surrounding metropolitan areas are expected to gain an additional 60 million residents between now and 2030, reinforcing the critical need to better coordinate land use planning with water availability.

MEDIA SUMMARY OF RESULTS:

1. The report cites four main water challenges, each of which, along with stricter conservation efforts, could be alleviated by less sprawl and more compact development to help ease the strain on existing systems:

- Old pipes—Rusting and dilapidated water infrastructure leaks away gallons and risks all-out collapse.
- High growth constraining supplies—Fast-growing regions cannot sustain current land use patterns or water use practices given projected population increases. "In infertile zones, days may be numbered for refashioning dusty scrub into suburban landscapes with expansive lawns and swimming pools."
- Contamination threats—Industrial chemicals and agricultural runoff permeate groundwater and settle into drinking sources. "We did a great job hooking everyone on clean up . . . but just scratched the surface on pollution sources."
- Failure to conserve—The nation's use of water has more than doubled since 1950, due to water waste, and to neglected leaks that drip 1.25 trillion gallons annually—the total consumption of Los Angeles, Miami and Chicago combined.

2. Of the 14 metro areas in the report, all but three—Minneapolis/St. Paul, Philadelphia, and Atlanta—have specific conservation programs in place, indicating that many local governments are actively seeking a change in consumer behavior. However, each of the areas also faces numerous challenges including old pipes, uncertain water supply and struggles with regional cooperation. Los Angeles was the only city cited as facing all three obstacles, making its water problems particularly urgent. Infrastructure 2010 holds up Australia as a model for water conservation, stormwater capture, and recycling, as well as more condensed land development practices, using a combination of basic and sophisticated techniques that could be applied in U.S. cities and others globally. (Residents pay $3.87 per cubic meter for water in Sydney; in Los Angeles, they pay $2.21.)

3. The report offers solutions to the nation's water infrastructure problems that are similar to those recommended for transportation issues, in that the solutions aim to foster collaboration among different governmental entities, incorporate land use planning into infrastructure planning, and accept higher user costs as a necessity. Among the "fixes" specific to water:

- Use federal allocations to encourage the creation of long-range regional management programs to integrate water supply and conservation strategies with population projections, agricultural needs and utility demand.
- Face reality, in that consumers and businesses will have to pay more to ensure reliability and safe supplies.
- Give top priority to repairing and upgrading existing systems.
- Incorporate land use into water management, including restricting develop-

ment in areas without ample future water resources; using only native spe-
cies in landscaping; building more compactly to reduce runoff and enhance
retention.

- Protect ecosystems to enable more natural storage and restoration.
- Use all available resources, including capturing rainwater, recycling waste-
  water, recharging groundwater, and making nonpotable water potable.
- Incentivize conservation by more closely linking water costs to system us-
  age, repair and maintenance.
- Invest in desalinization technology.

4. In addition to the research into water infrastructure systems, Infrastructure 2010
also tracks transportation investments made by the U.S. and countries overseas,
particularly in light of the recession. Despite the economic downturn, Europe and
Asia continued to invest hundreds of billions of dollars in infrastructure upgrades,
while U.S. efforts have remained more limited, primarily targeting existing work
revived by the economic stimulus package, and $8 billion for new high-speed rail
lines that, while significant by U.S. standards, is a small percentage of what is actually needed to build the rail projects. The report rates progress made over the past
year in various areas:

- Stimulus funds—used to fund existing stand-alone projects with minimal
  long term impact
- Forming a national infrastructure policy—dialog initiated, no policy en-
  acted or moving through legislative process
- Generating new revenues—no progress on raising gas taxes, user fees; too
  politically toxic
- Silo busting—signs of hope, as the U.S. Department of Housing and Ur-
  ban Development, U.S. Department of Transportation, and Environmental
  Protection Agency announce new Partnership for Sustainable Communi-
  ties
- New transportation funding legislation—stalled
- National infrastructure bank serving as private capital source—stalled

5. As in previous Infrastructure reports, Infrastructure 2010 urges a far greater
commitment by the U.S. "Economic fallout, competing priorities, and sticker
shock prevent the country from aggressively addressing a slow-motion meltdown,
the consequence of underinvesting for decades . . . In the meantime, inertia has
its own price. The more we let things go, the more expensive the costs to fix and
rebuild," the report states. "The world order now begins to pass America by as
countries in the European Union and Asia, in particular China, continue to implement policies to integrate rail, road, transit, airport and seaport networks to serve
major economic hubs, using 21st century technologies and systems."

6. A brief overview of infrastructure activity abroad—investments, major projects,
and challenges:

- China—China leapfrogs the rest of the world when it comes to building
  modern transport infrastructure, investing hundreds of billions of dollars
  in new roads, dams, mass transit, high-speed rail, ports and airports. The

government has directed most of $600 billion in stimulus funds to large-scale infrastructure, including nearly 10,000 miles of new high-speed rail to be completed by 2020.

- Japan—Japan has employed public works stimulus to boost its lackluster economy for two decades, building new roads and new airports, and expanding its "bullet" trains. Now it faces population losses and an aging population, leaving it with an overdeveloped infrastructure system offering more capacity than is warranted by demand.

- South Korea—In the country's pipeline: a 93-mile underground road network in Seoul budgeted at $9 billion, a $3 billion expansion of Incheon International Airport, $2.3 billion in green energy initiatives, and a $19 billion cleanup of major rivers, all demonstrating the country's ongoing commitment to advancing its infrastructure systems.

- Singapore—Singapore enhances is reputation for acclaimed infrastructure with the completion of the Marina Barrage, a $170 million hydroelectric dam project that integrates flood control, green technologies and recreation features.

- European Union—The EU provides $630 million to member nations to spend on rail links between countries, including high-speed lines. The push to create jobs advance infrastructure plans, but the short-term increased spending is likely to be followed by a slowdown to focus on the deficits created.

- United Kingdom—The UK adopts a combination of large-scale and small-scale infrastructure initiatives to reduce congestion around London. The 73-mile Crossrail tunnel will connect Heathrow Airport to the eastern suburbs; and outside the city, the Eurostar bullet train now extends to Amsterdam.

- France—France injected $1.21 billion in stimulus funds into its transport sector in 2009, and moves ahead with plans to double its high-speed rail system to 2,500 miles by 2020. The government also aims to be the world leader in developing infrastructure to support use of electric and hybrid electric cars.

- Germany—A consortium of German industrial, energy and finance companies pursues a $556 billion solar energy project to transport solar-generated electricity from state-of-the-art plants in the Sahara Desert to Germany and other European countries. It could supply as much as 15 percent of Europe's energy needs by 2050.

Bibliography

Courtesy of Brendan Hoffman/Getty Images for Amtrak

Visitors celebrate National Train Day at Union Station on May 7, 2011, in Washington, D.C.

Passengers board a subway train from a snow-covered platform at a Bronx stop on January 12, 2011, in New York City.

Books

ASCE Critical Infrastructure Guidance Task Committee. *Guiding Principles for the Nation's Critical Infrastructure*. Reston, Va.: American Society of Civil Engineers, 2009.

Ascher, Kate. *The Works: Anatomy of a City*. New York: Penguin, 2007.

Bluestone, Daniel M. *Buildings, Landscapes, and Memory*. New York: W.W. Norton & Co., 2011.

Brooks, Karl Boyd. *Public Power, Private Dams: The Hells Canyon High Dam Controversy*. Seattle, Wash.: University of Washington Press, 2006.

Brown, David J. *Bridges: Three Thousand Years of Defying Nature*. Richmond Hill, Ont.: Firefly Books, 2005.

Dellinger, Matt. *Interstate 69: The Unfinished History of the Last Great American Highway*. New York: Scribner, 2010.

Geddes, R. Richard. *The Road to Renewal: Private Investment in U.S. Transportation Infrastructure*. Lanham, Md.: AEI Press, 2011.

Glennon, Jerome Robert. *Unquenchable: America's Water Crisis and What to Do About It*. Washington, D.C.: Island Press, 2009.

Gutfreund, Owen D. *Twentieth-Century Sprawl: Highways and the Reshaping of the American Landscape*. New York: Oxford University Press, 2005.

Hayes, Brian. *Infrastructure: A Field Guide to the Industrial Landscape*. New York: W.W. Norton, 2005.

Huler, Scott. *On the Grid: A Plot of Land, an Average Neighborhood, and the Systems that Make Our World Work*. New York: Rodale Books, 2010.

Jones, David W. *Mass Motorization and Mass Transit: An American History and Policy Analysis*. Bloomington, Ind.: Indiana University Press, 2008.

Kasarda, John D. *Aerotropolis: The Way We'll Live Next*. New York: Farrar, Straus and Giroux, 2011.

Kennedy, Randy. *Subwayland: Adventures in the World Beneath New York*. New York: St. Martin's Griffin, 2004.

LePatner, Barry B. *Too Big To Fail: America's Failing Infrastructure and the Way Forward*. Lebanon, N.H.: Foster Publishing, 2010.

Lutz, Catherine. *Carjacked: The Culture of the Automobile and Its Effect on Our Lives*. New York: Palgrave MacMillan, 2010.

Malone, Laurence J. *Opening the West: Federal Internal Improvements Before 1860*. Westport, Conn.: Greenwood Press, 1998.

Marshall, Alex. *Beneath the Metropolis: The Secret Lives of Cities*. New York: Carroll & Graf Publishers, 2006.

——— . *How Cities Work: Suburbs, Sprawl, and the Roads Not Taken*. Austin, Tex.: University of Texas Press, 2000.

McNichol, Dan. *The Roads that Built America: The Incredible Story of the U.S. Interstate System*. New York: Sterling Publishing Co., Inc., 2005.

Motavalli, Jim. *Breaking Gridlock: Moving Toward Transportation that Works*. San Francisco, Cal.: Sierra Club Books, 2001.

Munson, Richard. *From Edison to Enron: The Business of Power and What It Means for the Future of Electricity*. Westport, Conn.: Praeger Publishers, 2005.

Nye, David E. *When the Lights Went Out: A History of Blackouts in America*. Cambridge, Mass.: MIT Press, 2010.

O'Toole, Randal. *Gridlock: Why We're Stuck in Traffic and What to Do About It*. Washington, D.C.: Cato Institute, 2010.

Rohaytn, Felix. *Bold Endeavors: How Our Government Built America, and Why It Must Rebuild Now*. New York: Simon & Schuster, 2009.

Schewe, Phillip. *The Grid: A Journey Through the Heart of Our Electrified World*. Washington, D.C.: Joseph Henry Press, 2006.

Schrag, Zachary M. *The Great Society Subway: A History of the Washington Metro*. Baltimore, Md.: The Johns Hopkins University Press, 2006.

Winston, Clifford. *Last Exit: Privatization and Deregulation of the U.S. Transportation System*. Washington, D.C.: Brookings Institution Press, 2010.

Web Sites

Readers seeking additional information on American infrastructure and related subjects may wish to consult the following Web sites, all of which were operational as of this writing.

American Infrastructure Magazine

http://americaninfrastructuremag.com/

With an emphasis on infrastructure case study projects, *American Infrastructure* is a national magazine produced by Peninsula Publishing that focuses its coverage on municipal trade and public works. Its subscribers include all levels of public officials as well as companies that do business with the government. Visitors to *American Infrastructure*'s Web site can read current and back issues of the magazine in full-color PDF format.

Report Card for America's Infrastructure

http://www.infrastructurereportcard.org/

Compiled every few years by the American Society of Civil Engineers (ASCE), the Report Card for America's Infrastructure provides a letter grade for each component of the country's vital public facilities, among them aviation, bridges, dams, and drinking water. The report card's multimedia Web site offers detailed examinations on the condition of each form of infrastructure and links to the data on which the letter grade was based. The material can be broken down by state as well as infrastructure category. Potential solutions to infrastructure shortcomings are also detailed. Founded in 1852, ASCE is the country's oldest national engineering organization; its membership is composed of over 140,000 civil engineering professionals.

U.S. Infrastructure Magazine

http://www.americainfra.com/

U.S. Infrastructure is a periodical published by GDS Publishing that covers "[t]he latest developments in American Infrastructure and Construction Management News," according to its Web site. On-line visitors can access a host of material, from features and interviews to blogs and news briefs. The site includes a wealth of multimedia content.

Additional Periodical Articles with Abstracts

More information about U.S. infrastructure and related topics can be found in the following articles. Readers interested in additional material may consult the *Readers' Guide to Periodical Literature* and other H.W. Wilson publications.

Tapped Out. Jennifer Grzeskowiak. *American City & County* v. 125 pp34–36+ May 2010.

At present, funding for water regulatory compliance is insufficient in many cases, and local governments are being forced to find new funds to keep water clean, the writer claims. The lack of money is due mainly to the combination of water infrastructure demands, low federal assistance, a strong municipal bond market, and a recession among residents that hinders increases in rates. Using private capital is seen as a feasible option for raising more funds. Alternative funding options, discussions with lawmakers, rate payer education, efficiency efforts, and financial responsibility are helping water authorities deal with the ever-changing regulatory environment. Clearly, there is no single solution, Grzeskowiak claims, but a common theme among almost every member of the water community is a sense of frustration with other stakeholders, from ratepayers to the federal government, who do not understand their work or its costs.

Rail Against the Machine. William S. Lind. *The American Conservative* pp22–25. v. 9 August 2010.

According to Lind, libertarians ignore the fact that highways only cover 58 percent of their costs from user fees, including the gas tax. To understand how conservatives might approach transportation issues more thoughtfully, he writes, we need to realize that all public transit is not created equal. You will find few people with alternatives sitting on buses crawling slowly down city streets, Lind contends. Most bus passengers, he writes, are "transit dependents" —people who have no other way to get around. But most conservatives have cars; they are "riders from choice," people who will only take transit that offers better conditions than driving. They demand high-quality transit, which usually means rail: commuter trains, subways, light rail, and streetcars. It is no coincidence, Lind writes, that the decline of America's cities accelerated when streetcars were replaced by buses. Streetcars are "pedestrian facilitators," he asserts, adding that it is easy to hop off, shop and have lunch, then get on the streetcar again when feet get tired. Pedestrians, Lind writes, are the lifeblood of cities.

Engine of Prosperity. Christopher B. Leinberger. *The American Conservative* v. 9 pp25–27 August 2010.

Transportation drives development, Leinberger writes, claiming that for the 6,000 years we have been building cities, the transportation system a society chose dictated what real estate developers could build. From ancient Sumer (present-day Iraq) to modern times, the transportation system is the rudder that steers the investment of a large portion of a society's wealth. So how, the author asks, do we pay for the transit, especially rail transit, that will allow developers to give the market what it wants: walkable urban development? The answer, Leinberger claims, can be found in the past. In the early 20th century, every American town over 5,000 people was served by a streetcar system. Until 1950, our grandparents did not need cars to get around because they could rely upon various forms of rail transit. The average household only spent 5 percent of its income on transportation 100 years ago, versus 24 percent for drivable households today. Leinberger writes that we were able to afford such an extensive rail system because real estate developers, sometimes aided by electric utilities, not only built the systems but paid rent to cities for right of way. According to Leinberger, they understood that transportation drives development and that development had to subsidize the transportation.

Reinventing How We Do Infrastructure. Heather Millar and Dan Smith. *American Forests* v. 115 pp34–40 Spring 2009.

Infrastructure dramatically influences the extent of the environmental damage that societies inflict, Millar and Smith write. Given this clear connection, they insist, the way in which people restore, expand, and manage these human systems will have repercussions not just for economies but also for ecosystems. In the middle of a global economic crisis, the quickening pace of environmental degradation, unprecedented spending to rebuild infrastructure, and a new administration in Washington, the writers renew the appeal by *American Forests* to keep nature fully in the picture.

Smart Grid Test Underwhelms. Paul Merrion. *Crain's Chicago Business* v. 34 p1 May 30, 2011.

Commonwealth Edison Co.'s first Chicago-area test of its new smart-grid technology did not connect very well with consumers, Merrion contends. A new report by independent researchers reveals that less than 9 percent of around 8,000 randomly chosen households used their newly installed smartmeters to save money by adjusting thermostats or switching off appliances during the afternoon or on hot days.

Infrastructure Needs—and Gets—Help. Scott Burns. *Earth* (Alexandria, Va.) v. 54 p86 May 2009.

According to Burns, geologists need to ensure that legislators are aware of the urgent need for improvements in the U.S. infrastructure. The fact that the infrastructure is aging, failing, and in desperate need of repair is highlighted by the

American Society of Civil Engineers' 2009 Infrastructure Report Card, which rates the overall infrastructure as a "D" and estimates that $2.2 trillion is needed to bring all categories to a passing grade. Congress and state legislatures have noted the need for improvement and have targeted large sums of money from recent economic stimulus efforts to infrastructure repair and upkeep. Geologists can help these efforts, Burns writes, by providing data on natural hazards risks for the planning process for new infrastructure.

Making Electric Grids Smarter. Brian Fisher Johnson. *Earth* (Alexandria, Va.) v. 54 pp36–43 May 2009.

Increasing investment in smart grid technologies is making the U.S. electrical infrastructure cleaner, more efficient, and more reliable, Johnson reports. Blackouts like the one that affected a large part of the American Northeast on August 14, 2003, highlighted the vulnerability of a system running at or near capacity, with little room for recovery in the event of a failure at an individual power plant. Although energy-saving appliances can play a significant role in reducing demand, Johnson contends, even greater improvements could be achieved through the use of smart grid technology. These use information technology and advanced transformers and meters to tell suppliers either when they can reduce electricity supply to certain households because of low demand or when to borrow electricity from those households to meet high demand elsewhere. The first fully integrated grid system in the U.S. is being installed in about 45,000 houses in Boulder, Colorado.

Digital Delivery. Dennis J. Wamsted. *Electric Perspectives* v. 36 pp20–24+ March/April 2011.

As utilities throughout the United States upgrade their distribution systems, Wamsted observes, it is evident that the transition is being pushed ahead significantly by the $3.4 billion in smart grid-related funding contained in the 2009 stimulus legislation. A McKinsey & Co. survey estimates the potential market for utility smart meter applications at $59 billion. Wamsted discusses the digital transition that is under way in the utility arena and the benefits that it is likely to bring about.

Solutions for America's Infrastructure Crisis. *Governing* v. 24 pp 51–54 November 2010.

According to the author, the United States is seriously underfunding infrastructure, and, as a consequence, safety, quality of life, and economic prosperity are being sacrificed. A recent series of *Governing* editorial roundtables was held in Houston; Sacramento, California; Boston; Raleigh, North Carolina; and Omaha, Nebraska; to discuss the chronic neglect and underfunding of critical infrastructure. It was sponsored by the American Society of Civil Engineers, whose president, Blaine Leonard, explained that the roundtables were an attempt to highlight the problem and gather some of the finest minds from government and the private sector to resolve the nation's intensifying infrastructure crisis.

'L' Capitan. Zach Patton. *Governing* v. 23 pp20–23+ July 2010.

Richard Rodriguez has run the Chicago Transit Authority (CTA) since March 2009 and has garnered a reputation for picking up trash whenever he rides a bus or rail car in the city, Patton writes. Although Rodriguez is focused on the details, Patton observes, he simultaneously balances many bigger issues, and they are much thornier than errant candy wrappers. The CTA is coming out of the worst budget crisis of its 60-year history. Huge gaps in funding have forced sweeping layoffs and service reductions. Furthermore, the system, some parts of which date to the 1890s, requires billions of dollars of infrastructure repairs and upgrades.

Not So Fast. Josh Goodman. *Governing* v. 23 pp20–24+ May 2010.

As California advances with perhaps the country's most ambitious high-speed rail project, and, according to Californians, the largest public works project in U.S. history, the uncertain fate of El Palo Alto, the famous 1,070-year-old redwood tree that stands only a few feet from the proposed train tracks, is just one indication of the complexities of building 800 miles of new infrastructure in a heavily developed, densely populated state, Goodman writes. Goodman considers a number of questions, such as how a state pays for such a system, who operates it, where the tracks and stations should be placed, and how to minimize disruptions to the environment and to communities that suddenly will have trains speeding through them at up to 220 miles per hour.

Stuck in the Devil's Triangle. Donald F. Kettl. *Governing* v. 23 pp14–15 April 2010.

The American Society of Civil Engineers (ASCE) estimates that the country faces a $2.2 trillion infrastructure backlog, Kettl observes. One of every eight bridges is "structurally deficient," and 85 percent of public transit systems are struggling to carry the increasing number of riders. In Kettl's view, the problem is a devil's triangle of cross-pressures that make finding a solution incredibly difficult. On one side of the triangle is the deep and continuing state budget crisis. Job recovery from the Great Recession is slow, as is the recovery in state revenues. Medicaid is still draining state budgets, and the hemorrhage is increasing as aging baby boomers start drifting into government-funded nursing homes.

A Bridge to Somewhere. Alex Marshall. *Governing* v. 22 pp22–23 June 2009.

Instead of reacting to demand, transport planners should anticipate demand, Marshall contends. Building a bridge, road, or an airport that leads to depressed areas is riskier than merely responding to demand, he admits, but the payoff is potentially larger. Americans appear to recognize that infrastructure investment can jump-start an economy, but as states and localities decide what to fund, they should not simply produce studies to show where the traffic is, Marshall adds. Rather, they should question how transportation can boost commerce and trade and make better places to live.

The Incrementalists. Karrie Jacobs. *Metropolis* v. 30 pp42+ January 2011.

The problem with the United States' current approach to infrastructure is that it is funding a grab bag of projects, rather than constructing any kind of network, Jacobs states. According to Jacobs, if the country were truly serious about reducing its use of fossil fuels and transforming how Americans travel, it would be constructing something exactly as large as the interstate highway system. While the United States may not have the money at present, she notes, the real problem is that the federal government is scared to act like a federal government: top-down and overbearing.

The Road Ahead for Infrastructure. Sibley Fleming. *National Real Estate Investor* v. 51 p16 May 2009.

A new report from the Urban Land Institute and Ernst & Young maintains that the American Recovery and Reinvestment Act does not do enough to fix the nation's failing infrastructure, Fleming observes. According to the report, "Infrastructure 2009: A Pivot Point," the $132.4 billion allocated to repair and maintain roads, bridges, and water supplies is only a fraction of the $2.2 trillion required. Fleming provides detail of the report's findings.

If You Build It . . . Ezra Klein. *Newsweek* v. 156 p21 October 11, 2010.

According to Klein, the Obama administration's plan to invest in the nation's infrastructure is too modest. The announced $50 billion increase in infrastructure spending is only for surface transportation, but the water system, schools, and levees should also receive attention, he writes. The recession offers a rare opportunity to do what is necessary and to save money in the process, Klein maintains, adding that construction materials and labor are cheap and borrowing costs are low so every dollar of investment today will go much further than it would have five years ago or than it is likely to stretch in five years time.

Prosperity Depends on Infrastructure. Barry B. LePatner. *Outsourced Logistics* v. 2 pp16–17 March 2009.

Even as U.S. policymakers and pundits focus on the financial crisis affecting America, a further crisis is developing in the form of a deteriorating infrastructure system, LePatner observes. For example, a recent study reveals that the United States has experienced more than 500 bridge failures since 1989. Although it is clear that America's infrastructure shortcomings cannot be fixed overnight, LePatner notes, moving aggressively toward a solution now rather than applying a series of ineffective cosmetic measure would result in real improvements that would benefit the country in the long term. Addressing critical transportation and infrastructure issues will require a national commitment and a strategic plan that should include various solutions, the writer contends, going on to discuss several of them.

Falling Down. *PM Network* v. 24 pp16–17 February 2010.

Despite current spending on infrastructure as part of stimulus plans, increased traffic and lack of maintenance have left the world's bridges in crisis, the writer contends. However, some progress is being made in major renovations and new construction. Project managers involved in such projects can expect a high level of scrutiny because of the high-profile nature of the work.

Failure Notice. *PM Network* v. 24 pp32–33 February 2010.

According to the American Society of Civil Engineers (ASCE), it would take five years and $2.2 trillion to fix the deficiencies of U.S. infrastructure. In ASCE's 2009 Report Card for America's Infrastructure, transportation, energy, public facilities, water, and waste averaged a D grade, with no category receiving a grade higher than C+ and only energy, which went from D to D+, having improved since 2005.

Full Speed Ahead. *PM Network* v. 23 p25 June 2009.

In an interview, Superior Court judge Quentin Kopp, chair of the Sacramento-based California High-Speed Rail Authority, discusses his role in planning, constructing, and operating an 800-mile (1,287-kilometer) train system to serve the state's major metropolitan areas. Kopp was instrumental in getting California voters to pass a $9.95 billion bond to fund the 20-year project and has instituted rigorous government oversight to ensure that it remains on track.

Renovating America. Adam M. Bright. *Popular Science* v. 276 pp38–45 February 2010.

The writer suggests a number of technology ideas that can transform American infrastructure. The ideas are sorted out across five categories—transportation, water, power, telecom, and sewage.

Water Infrastructure in Crisis. Marc Santora and Rande Wilson. *Public Management* v. 90 pp17–20 December 2008.

According to the authors, the integrity of America's drinking water and wastewater infrastructure requires a concerted effort to increase the sustainability of the sector's critical assets. The Local Government Advisory Committee of the Environmental Protection Agency (EPA) notes that if investment in this infrastructure is not increased, the funding gap could reach $225 billion by 2020. The problem is that artificially low rates have contributed to a serious deterioration of essential infrastructure, making full-cost pricing increasingly important to the water sector. Moreover, the safety, security, and resiliency of municipal water supplies and clean water programs need adequate resources, which makes the incorporation of risk management into asset management vital. Santora and Wilson offer local government managers advice on achieving this.

Stimulus Success Story. George Allan. *Public Works* v. 141 pp48–50 December 2010.

Allan highlights the $2.88 million Water Treatment Plant Solar Photovoltaic Array project of the Massachusetts' Chelmsford Water District. Using stimulus funds from the American Recovery and Reinvestment Act (ARRA) of 2009, Massachusetts's goal is to improve solar capacity at public plants by over 132 megawatts. Currently, the district's treatment facility in Crooked Spring is one of New England's largest photovoltaic arrays. The recently completed 485-kilowatt solar field is expected to provide approximately 55 percent of the facility's annual power demands, Allan observes.

Get Ready to Work with the Railroad. Stephanie Johnston. *Public Works* v. 141 p7 November 2010.

The writer reports on how Americans passively fund improvements for infrastructure programs in the United States. One example is the American Recovery and Reinvestment Act of 2009 which allocates $8.8 billion for high-speed rail programs.

The Looming Crisis in Transportation: Heartland Corridor. William C. Vantuono. *Railway Age* v. 208 pp26–28 March 2007.

With freight service exceeding the country's ability to supply it, experts predict that U.S. rail infrastructure will face grave danger in the near future, the writer reports. Experts agree that the privately funded, self-capitalized railroads are best equipped to meet this crisis head-on. However, they will require help in doing so in the form of significant investment through public/private partnerships (PPPs). Vantuomo goes on to discuss the use of PPPs in addressing the looming U.S. transportation crisis.

Mystery Train: California's High-Speed Rail Project. Tim Cavanaugh. *Reason* v. 42 pp82–83 August/September 2010.

California's 14-year-old high-speed rail project typifies waste, Cavanaugh asserts. Since 1996, the bullet-train project has cost taxpayers over $250 million, but no track has yet been laid. A recent report by the state's auditor, Elaine M. Howle, spells out the free-spending ways of the California High Speed Rail Authority, which, among other things, seems to have done no research on the project's potential revenue and paid at least $4 million of invoices for which it had no evidence that the contractors carried out the work for which they paid, Cavanaugh states. Moreover, California's legislative analyst's office had already noted that the authority's plan contains no timeline and no specifics and seems to violate the law by using bond funds to subsidize operations, the writer adds.

Farewell to the Drive-in Utopia. James Howard Kuntzler. *Salmagundi* pp82–96 Fall 2010/Winter 2011.

The writer discusses how American-style suburbia embodies a living arrangement with no future. He believes that American-style suburbia is "the greatest misallocation of resources in the history of the world." This, he asserts, is apparent because the country will not be able to sustain the great project of suburbia in a few years into the future as the country's oil supply gets more limited, and one has to face the disappointing reality that the ostensible alternative energy will not come close to counteracting the oil losses. Kuntzler contends that investing America's enormous accumulated capital treasure into building an infrastructure for everyday life with no future is a tragic misallocation of resources. He also notes that there is no previous case in history of a civilization making capital investments as carelessly as the United States has and establishing what seems to be a throwaway human habitat.

He affirms that now suburbia's time is over, and the aspects of suburbia that already exist will increasingly lose value and utility.

A Tale of Two Flows. Walter Schwarz. *Water Environment & Technology* v. 22 pp32–35 November 2010.

The City of Fort Lauderdale in Florida has collaborated with Colorado's CH2M Hill to implement a WaterWorks 2011 project to overhaul the city's water, wastewater, and sewer infrastructure, Schwarz reports. Even though the $740 million city-wide program management initiative focused mostly on new construction, the city has invested $35 million in trenchless technology to repair the existing collection system. This ongoing sewer rehabilitation effort is expected to reduce infiltration and inflow (I/I), thereby reducing base flows and peak flows at the G.T. Lohmeyer Wastewater Treatment Plant. The results showed that the sewer rehabilitation project has reduced both groundwater I/I and rainfall-derived I/I levels, thereby increasing plant capacity.

Whose Road Is It, Anyway? Rebecca J. Rosen. *The Wilson Quarterly* v. 32 pp50–51 Spring 2008.

In this article, part of a special section on American infrastructure, Rosen writes that state and local governments are confronting serious costs for maintaining and improving infrastructure. The U.S. Chamber of Commerce forecasts that, by 2015, there will be a shortfall of $1 trillion in investment, with many states facing daunting deficits. A significant factor in the crisis in financing both the maintenance and expansion of transportation infrastructure at the state level is the fall in revenue from motor fuel taxes, prompting states to seek creative ways to fund their transportation needs. Some states are working with private firms to develop dedicated lanes on busy roads, which use variable tolls to control volume and permit cars to maintain higher speeds.

Index

About the Editor

A Connecticut native, **PAUL MCCAFFREY** was born in Danbury and raised in Brookfield. He graduated from the Millbrook School and Vassar College in Dutchess County, New York, and began his career with the H.W. Wilson Company in 2003 as a staff writer for *Current Biography*. He has worked on The Reference Shelf series since 2005, personally editing a number of titles, among them *The News and Its Future*, *Hispanic Americans*, *Global Climate Change*, and *The United States Election System*. As a freelance author, he has written several biographies for Chelsea House. He lives in Brooklyn, New York.